The Owner Builder

Leeza Sipek, Matthew Sipek and Hazel Foote

PUBLISHING

National Library of Australia Cataloguing-in-Publication entry

Sipek, Leeza

The owner builder/by Leeza Sipek, Michael Sipek and Hazel Foote.

9780643100428 (pbk.)
9780643106970 (epdf)
9780643106987 (epub)

Owner-built houses – Australia – Amateurs' manuals.
House construction – Australia – Amateurs' manuals.
Dwellings – Maintenance and repair – Amateurs' manuals.

Sipek, Michael.
Foote, Hazel.

690.80994

Published by

CSIRO PUBLISHING
36 Gardiner Road, Clayton VIC 3168
Private Bag 10, Clayton South VIC 3169
Australia

Telephone: [+613] 9545 8555
Local call: 1300 788 000 (Australia only)
Fax: +61 3 9662 7555
Email: csiropublishing@csiro.au
Web site: www.publishing.csiro.au

All photographs are by the authors.

Set in Adobe Minion Pro 11/13.5 and Adobe Helvetica Neue LT
Edited by Elaine Cochrane
Cover and text design by James Kelly
Typeset by Desktop Concepts Pty Ltd, Melbourne
Printed by Ingram Lightning Source

Feb26_RP_ILS

Contents

1
Introduction

Welcome to *The Owner Builder*. Our mission is to provide an easy-to-understand plain English building guide for you and all those other creative Australians who want a better financial future and lifestyle with their families.

Who is an owner builder?

Owner builders are not in the business of building. When people say they intend to be an 'owner builder', this may mean different things to different people.[1]

1. You take on the role of the builder and build everything yourself without engaging tradespeople except in areas where licensed tradespeople are required by law, such as for plumbing and electrical work.
2. You are the builder and do some of the work yourself. You will oversee the project through to completion, but will hire subcontractors or tradespeople to do part of the building work, for example framing or roofing.
3. You act as the construction manager. You organise all the materials and subcontractors including perhaps a registered building contractor to create your home. You carry insurances, organise site management, and organise inspections of the progress of works.[2]

Managing a development project doesn't take an expert craftsman, but a person who can schedule, delegate and direct traffic on the construction site. By taking on the role of a builder you will need to be confident that you have the time and ability to undertake these tasks. You will need to be able to manage and supervise your contractors. You must also have a good understanding of the tasks they perform and the materials being used.

Why owner-build?

Currently there are around 40 000 owner builders taking on their own construction projects nationally each year. Some say that as much as 40% of the cost to build can be saved by owner builders. Of course, whether you can translate these savings to your project depends on how competent you are in carrying out the many and varied tasks involved.

There are five main reasons to undertake your owner builder project:

1. To manage a smaller mortgage.
2. To save an average 30% of a builder's costs.
3. To be in control of your time and money.
4. To ensure the best quality workmanship.
5. To create a custom design.

It is important to achieve the outcomes sought. This book can help you, the owner builder, gain confidence, and provide you with support in taking on a building commitment. It shows by example and advice how, just by being the organiser of the project, you can save thousands of dollars.

If you decide to hire a contract builder you can benefit from this guide, as it helps in understanding what is involved in building a home and what to expect in the process. It has a personal sequential photo gallery of the building process in sixty-eight steps. It includes tips and hints on the approval process, internal and external decorations, construction ideas and, most importantly, shows how to avoid mistakes.

Vital information includes aspects of choosing a block of land, such as being careful with slope on the site and determining the foundations and retaining walls required. These basics will have a lot to contribute to the outcome of the project and assist in selecting a type of house design that suits your land and budget. In general this book will aid the decision-making whether you are undertaking owner building, contract building, renovating, or hiring a project manager to build the home on your behalf.

This book eliminates the need for you as owner builder to attend long-term building courses (although you may still be required to undertake a short owner builder course to obtain your building permit). It spells out all necessary requirements and provides a straightforward guide for owner builders with little or no knowledge or experience in building.

As authors of *The Owner Builder* we have over 17 years combined project building industry experience. We have used our experience in sales consulting, management and owner-building to assist colleagues and the public with building ideas, problem-solving, product searches, and differentiating between dream goals and reality goals. Structure and budgeting was often a prime concern of those who sought help. This knowledge and advice is now available to you in this book.

The photographer has a wide range of knowledge of the commercial construction industry and owner-building. His input to the text of *The Owner Builder* as well as the easy-to-understand construction pictures adds to the readability and simplicity of the instructions. Using the methods and outlines contained within this book with similar types of materials and construction processes will enhance your chances of a successful completion of the dwelling and obtaining the final occupancy certificate.

This guide to the owner builder outlines the responsibilities that you need to be undertake and explains how to put your plans into action. It shows each step of construction on site in the building process, including the early stages of planning, and the essentials of design, materials and land qualities you need to look at to stay within a budget.

It gives examples of experiences on site and tips on dealing with tradespeople. It lists and explains necessary insurances to purchase, and the contractors' insurance and warranties required by law. Steps to take control of the budget are outlined, and the value of hiring an estimator to prepare a tender is examined. Building materials that are mandatory for meeting sustainability requirements (such as BASIX in New South Wales) are discussed. Bushfire safety and protection are crucial subjects. The option of purchasing an existing house to knock down and rebuild is assessed.

The value of having professional legal advice and taking precautions when signing contracts is emphasised, and advice is given as to whom to turn to when a dispute arises. Suggestions for achieving a resolution are given.

Theft from building sites is a well known hazard, so advice in loss-prevention is provided.

The choice of private certifiers or council building surveyor inspections and weighing up the difference is another topic examined, as it depends on time and financial constraints, as does the choice of contract building versus owner-building.

The exciting part of deciding on colours and concepts can also be challenging so, with the benefit of their experience, the authors provide some ideas and concepts to choose from and some problems to avoid.

Landscaping must-haves to achieve the famous Aussie backyard and incorporate individual needs are often best planned ahead and prepared for during the building process, and this guide will outline when such steps need to be taken.

The appendices provide comprehensive lists resources such as websites, owner builder centres and other relevant information for tradespeople, materials suppliers and services.

Are you up to it?

The role of an owner builder is demanding, especially if you have a full-time job and a family. As an owner builder you have decided to take on all the tasks that a building contractor must do to complete a successful construction project. You must allocate time to visit the site and keep activities to schedule, check occupational health and safety requirements, order materials, visit suppliers and hardware stores, and make phone calls to remind tradespeople to advise you of delays or revised site timing schedules.[3] Your dream home can become a nightmare if you do not plan correctly.[4]

In making a decision to commit to owner-building, you need to consider the following questions:

- Are you good at managing situations?
- Are you good at organising people?
- Do you think ahead and plan?
- Do you use common sense?
- Do you have the time to commit to the project?
- What is your knowledge of the building industry?
- What is the condition of your health?
- Are you able to handle financial or contractual disputes with subcontractors?
- What about your availability to be on-site to receive materials and ensure that they comply with specifications of required quantity and quality?
- Do you have the ability to distinguish technically what is defective building work?[5]

Once you have explored and absorbed the contents in this book, owner-building will become a very rewarding and achievable challenge.

2

Responsibilities of the owner builder

It's the law to obtain all building approvals

The owner builder needs to be aware of legislative requirements. This is why some state authorities have introduced an owner builder course: to enable the owner builder to prepare, plan and manage building projects. Once the development application has been approved, there will be a list of conditions of consent to obey. We suggest you take the time to read the Building Code of Australia for a better understanding of the Australian Standards. The Building Code of Australia requires:

- all required inspections
- a structurally sound dwelling and retaining walls
- safety for the environment
- minimisation of impact on the neighbourhood
- energy efficiency housing.

Council and authority approvals

Owner builders have a better chance than contract builders of receiving approvals quickly. Our experience working for contract builders has taught us that the administration staff of contract builders are very busy handling multiple council applications. As an owner builder you are usually able to monitor the progress of your application with council more closely.

Some other reasons for the delays with contract builders are:

- limited cash flow in the company
- large numbers of house orders
- variations of working drawings
- builders' employees not working efficiently
- errors in applications and paperwork.

The approvals you need

The following are mandatory requirements:

- white card safety induction course
- owner builder course (required in New South Wales, ACT and Queensland)
- building permit
- finance
- sustainability standards (for example BASIX and NatHERS Certificate approvals in New South Wales)
- developer application (if the developer selling the land has guidelines that must be followed)
- council approval
- water board office and inspections
- termite protection inspection
- structural engineer piers and slab inspections
- private certifier for the commence construction certificate or certificate of compliance
- private certifier for the occupancy certificate.

Depending on the proposed dwelling and the location, you may also require the following:

- additional council requirements as specified in the development control plan
- hydraulics engineer
- shadow diagrams and landscape plans (for drawings)
- mine subsidence board
- demolition of existing dwelling application
- heritage codes on existing buildings
- bushfire plan site assessment
- termite protection ant caps on stumps if using physical barriers.

Each council has its own town planning requirements and protocols. As an owner builder you need to request the relevant development control plan (DCP) from the council for the area in which you wish to build. There may be even more requirements, such as environmental planning, heritage and assessment regulations to follow. The council town planners are there to assist the ratepayers with all building inquiries and requirements, so owner builders are well advised to take advantage of their services.

Options available

If you are a creative person and prefer the actual construction stages of the project rather than the approval and paperwork stages, you have the option to hiring a consultant service to manage the approval process. We used this type of service for our own project and it wasn't very expensive. You can find businesses offering these services through the Yellow Pages, via the internet, or in any advertising media related to project building.

State authorities

The following state authorities are responsible for issuing permits and approvals:

- ACT: Office of Fair Trading <www.ors.act.gov.au/FairTrading/index.html>
- New South Wales: Department of Fair Trading <www.fairtrading.nsw.gov.au>
- Northern Territory: Consumer Affairs <www.nt.gov.au>
- Queensland: Office of Fair Trading <www.fairtrading.qld.gov.au>
- South Australia: Office of Consumer and Business Affairs <www.ocba.sa.gov.au>
- Tasmania: Office of Consumer Affairs and Fair Trading <www.consumer.tas.gov.au>
- Victora: Consumer Affairs <www.consumer.vic.gov.au>
- Western Australia: Department of Commerce <www.commerce.wa.gov.au>

Private certifiers

A private certifier is an accredited person who is suitably qualified to inspect and provide documentary evidence of compliance of construction as detailed in the Building Code of Australia. Our private certifier inspected our plans to ensure they were drawn according to regulations, prepared registration certificates with council, inspections, the construction certificate and the occupancy certificate. Our private certifier came to the site for an inspection of works after the following four stages:

- frame, steel works, footings, floors, walls and concrete poured (inspection by council or private certifier)
- wet area, waterproofing (bathroom, shower, bath) and flashing (between roof and tiles) (inspection)
- termite control around the building (termite company inspection)
- final: building completed and before occupation (inspection).

There were also in-between approvals and additional paperwork as follows:

- contour levels
- draftsman (to design working drawings)
- colour selections (for kitchen doors, paint etc.)
- insurances (for workers on site etc.)

- geotechnical study (on the land)
- surveyor and peg out of dwelling (by surveyor before the concrete slab is poured).

How long do approvals take?

The average time taken for approvals is six months. This estimate is based on simple proposals and councils that are easy to deal with. It is difficult to understand why this takes so long. However, it is a fact the owner builder needs to confront, especially if you are renting while taking on a building project. It is important to monitor the approval process. For every department that you are waiting on, the idea is to have direct contact if possible with the department or person dealing with the application. Ask for an approximate time frame for the application. If it goes over that time frame, you can question the delay.

To take into consideration and understand the volume of applications departments are assessing, we need to look at where large growth areas are located and where vacant land is available. Sometimes there is a need to travel to the proposed site for an inspection to complete the application approval. Also, there are the usual office delays such as lunch breaks, annual leave, meetings, short staff and sick leave. At other times there is no given reason for why so much time was taken to approve the application. It is essential to remain cool, calm and realistic to avoid any disputes arising.

As an owner builder you need to monitor the progress of your application with regular calls to authorities for status updates. This may help you receive the application approval more quickly. The owner builder course is there to help you understand the requirements for owner-building and is highly recommended.

Supervising subcontractors and services

As an owner builder you are required to supervise all tradespeople, business people and anyone else involved in the construction. There are legal requirements and liabilities when you are committing to a project.

Grounds for dissatisfaction with a particular trade, business or professional service on site include tradespeople:

- not fulfilling a scheduled job according to contract or plans
- under the influence of drugs, including alcohol
- not punctual
- bringing animals or pets on site
- not licensed or insured as required
- not legally entitled to work in Australia
- not keeping the worksite in a reasonable state.

The time frame for our approvals

White card and owner builder course: 3 days
Building permit: within minutes (at Department of Commerce)
Finance pre-approval: within 1 hour (1–2 weeks for secure finance with credit checks etc.)
Sustainability standards (BASIX or NatHERS approval): 1–2 weeks
Developer application: 2 days
Council approval: 6 weeks
Water board office: 3 weeks
Private certifier: 4 weeks
Engineer for plans: 2 weeks
Finance – release of funds: 1.5 weeks
Electrician certificate: end of electrical works
Termite protection inspection: 1 week

The state authority that issues building permits will be able to provide you with a list of relevant responsibilities, as they differ slightly from state to state. Please remember, the owner builder is classed as the builder of the project and not the building surveyor. The authority can also help with any disputes that may arise and can give advice if there is a problem or difference of opinion with an encounter.

What work can the owner builder do?

It is necessary to learn the labour limitations of owner-building. While there is the potential to do some work on site to save money, specialist work needs to be completed by a licensed tradesperson. The following are some examples that require licensed tradespeople, but can vary from state to state:

- electrical
- plumbing
- air-conditioning
- refrigeration
- gas-fitting work
- bricklaying
- carpentry
- concreting
- excavating
- fencing
- roof work
- structural landscaping

Work the authors actively took on

- Cleaning of site and removal of debris
- Labouring for bricklayer
- Labouring for roofer
- Painting internal walls, doors, skirting and architraves
- Wall insulation
- Bagging the exterior of the house and garage

The opportunity to physically get involved in the project gave them much satisfaction.

- piering
- wall and floor tiling
- waterproofing
- wet and dry plastering
- general building
- kitchen and bathroom renovations
- demolition
- erection of prefabricated metal home additions
- flooring
- building consultancy.

In addition, residential work needs to be completed by a licensed tradesperson and/or needs a building permit where the cost in labour and material exceeds a certain amount. The cut-off figure varies from state to state; it is $1000 in New South Wales. Works may be done without a permit if they are non-structural only.

A safe working environment

Occupational health and safety regulations protect workers and other people, and business services. They ensure correct disposal of hazardous materials and poisons, and limit noise. A major concern for the owner builder is to ensure the

Our experience with OH&S

One experience of safety issues on site we had relates to our first bricklayer. He would walk around on site and lay bricks wearing no shoes. He explained to us that he didn't like wearing shoes, but, as we were responsible if anything happened, we showed our concern and encouraged him for his own protection to wear shoes in accordance with occupational health and safety regulations.

building site is safe. This includes preventing falling materials and the wearing of protective clothing. Unauthorised people need to be excluded from unsafe areas.

Most states have mandatory regulations requiring owner builders to attend or complete owner builder safety-induction courses. These cover important guidelines for occupational health and safety (OH&S) and practical advice on implementing them.

Routine safety measures to be implemented include:

- cleaning the site regularly
- providing a safety rail for roofers
- council laws, e.g. demolished buildings and insurance
- providing scaffolding for bricklayers
- temporary fencing to exclude unauthorised people
- not leaving live electrical wires unattended
- tagging and testing all trade equipment in accordance with OH&S legislation.

Licensed contractors

A licensed contractor is a person who has formal permission to contract work and supply at a quoted price and is registered with a state authority in Australia.

It is important to contact the relevant state authority before hiring any contractors to check if the person is licensed and has a registered business. Ultimately, the license will protect you against inadequate workmanship and poor business practices. Usually, the contractor being considered for the job will provide a business card with the information and registration number required to do the license check. Following appropriate guidelines for license checks will result in fewer disputes and introduce competence and professionalism in business transactions and services.

Building permits and building disputes may also be handled by some of the above departments (see Chapters 13 and 14).

State departments providing licensing checks on contractors

ACT: Australia Capital Territory Planning and Land Authority
New South Wales: Department of Fair Trading
Northern Territory: Building Practitioner Board
Queensland: Building Services Authority and Q CAT
South Australia: Office of Consumer and Business Affairs
Tasmania: Worker Place Standards Tasmania
Victoria: Building and Licensing Commission
Western Australia: Builders Registration Board of WA

An unreliable tiler

When we were building our second house, the floor-and-wall tiler scheduled to work on our house did not turn up, even though he had confirmed the day before that he would definitely be there. His explanation – given later – was that he had to go to another job for a week. We simply cancelled his services and looked for another tiler.

This cancellation interfered with all the other tradespeople and professionals we had pre-booked for certain dates, and it affected their work flow and commitments to other clients.

The delay it caused to our project was two months, as we had to adjust to new dates for the kitchen installation, vanity, toilets, showers, laundry tubs and cupboards. The reason for such a long delay was that tradespeople and professionals book up work weeks and months ahead of time.

We used the time to continue with work we could do ourselves, such as: painting doors and architraves.

Managing the site and materials

The aim of managing the site and materials is to schedule tradespeople and arrange for delivery of materials in the most efficient way to avoid wasting time and money.

Methods of organisation can vary. However, our experience shows it is never too early to talk to and organise trades and professionals for the proposed dwelling. To ensure a successful project and limit time wastage, keep in touch with the tradesperson/business as the need for their services is approaching. You will need to think ahead and plan the next stage of building.

Occasionally, some people will be unreliable and not turn up on site. We recommend phone contact with tradespeople every two weeks to confirm they will be available for your project. This inconvenience can interrupt the flow of the project. If this occurs, contact the trade/supplier to determine the reason for their absence and judge whether to rebook or search for a new trade supplier.

Insurance

Insurance is the purchasing of a policy from an underwriter of insurance against loss or harm to persons or property, with an agreement of cost or refund between both parties.

Local councils and financial lending institutes require you to purchase insurance before you commence any works on site. To ensure you have appropriate insurance cover it is advisable to consult an experienced owner builder insurance broker who can tailor the insurance to suit your project.

The areas of insurance covered in comprehensive policies for owner builders are public liability, workers compensation, home owners warranty insurance, fire, storm, theft and vandalism. Note that an owner builder is still responsible for any future faults, even after the home warranty has expired, if the workmanship was not acceptable at the time of the building stage. See Chapter 9 for more details on insurance.

3

Obtaining finance

Most financial lenders understand the need to accommodate owner builders and there is a market for them, but it is ideal if you approach a mortgage broker experienced in owner builder loans. It is considerably easier to obtain finance if applicants have some equity and are able to prove the payment of past liabilities and consistent savings.

Financial lenders have long seen owner builders as a high risk. This is because the owner builder makes multiple purchases, has little room for error, and often has no experience in controlling a building project. To enhance your chance of obtaining a loan you will need a fully itemised tender or quotation to present to the lender. If the tender doesn't allow for certain items, the lender won't allow money for those items in the loan. Discuss options available with the type of loan you apply for.

The value of a tender

We did not manage the first project we owner-built very well. We did not have a tender and did not borrow enough funds for the entire project. We already had a lot of equity for the land and had asked for $180 000 to cover owner-building the dwelling and external essentials. The lender allowed this without a tender, and we found out well into the project that we were short by $40 000. Our misjudgements made it difficult for us to draw more funds from the lender until our mortgage broker interceded and reassured with the loan assessor about our job security. We then received the additional funds to complete the works.

What do lenders look for?

Credit history

Even the simplest commitment, such as to a phone company, car loan or credit card, that has outstanding arrears can result a loan application being rejected. The final outcome of the loan assessment will depend on what the default is and the value.

Liabilities and assets

An applicant's borrowing capacity is determined by comparing income and liabilities. Spending habits and reliability in making repayments are considered. Assets and collateral can be anything of value the applicant may own and are a form of security for the lender if the borrower cannot make repayments or the needs to sell the property. Collateral may include such items as cars, jewellery, furniture, personal items of value and property. The security and the purpose of the loan or item would also be taken into consideration by the lender. Mortgage insurance is generally required for low equity applicants. The majority of finance lenders will waive mortgage insurance when there is a minimum of 20% equity in the property of interest.

Security of repaying the debt

The lender will need to decide if the applicant can service (pay back) the loan when other liabilities and commitments have been allowed for. A letter of employment from the applicant's employer and pay slips will usually be required. These will provide details of the period employed, hours per week, gross income figures, hourly rate, tax rates, performance of employment and service, sick leave and holidays available. Basically, the lender wishes to assess the applicant's job security and responsibility.

Savings history

The size of the deposit the applicant has saved will help determine the outcome of the loan application and mortgage contract guidelines. Lenders are inclined to charge higher interest rates to applicants with no or very little deposit as opposed to someone with 10% or more deposit. In other words the more deposit the applicant has accumulated the better bargaining power the applicant has in negotiating the loan, as there is less risk to the lender. The strength of your ability to repay a home loan can be demonstrated by regular savings deposits from your salary or earnings of amounts similar to the future loan repayments over at least a six-month period. The history of other major purchases can also help when applying for loans.

Even holidays count

Before I applied for a home loan to purchase an existing apartment, I paid regular payments to my travel agent for a holiday over a twelve month period. As I was a 23-year-old woman with a single income the lender was concerned about the application. However, the copy of regular payments for my holiday got me over the line for a loan approval. Taking out income protection insurance also helped with my application.

Line of credit

A line of credit is the most convenient type of loan. It is like having money in the bank ready to pay your tradespeople and suppliers. If you plan to operate under a line of credit loan:

- stick to your budget
- manage all outgoings yourself
- keep track of estimated tenders or quotes
- only choose this system if you are good at controlling money.

A line of credit is usually approved to people that have 20% or more equity. The best way to apply for a line-of-credit is to state that the money is for 'future investments'.

Australian Tax Office inspections

The Australian Tax Office (ATO) spasmodically investigates owner builders' license and permit registrations for the payments between owner builders and the tradespeople they hire. There is evidence of some tradespeople operating on cash payments and leaving owner builders at risk of loss of insurance protection and warranties.

Keeping receipts and records of transactions, including contracts between the parties, will avoid the fraud of tradespeople not declaring their full income and

How we obtained finance

The assistance of a mortgage broker and his loan-writing approach to the lender made gaining finance for our second house go smoothly. The main reason given to the lender for the loan was 'The funds will be used for future investments'. We were granted a line of credit up to $250 000.

protect your investments. Electronic methods of money management and documentation such as direct deposit, credit card, eftpos and cheques can assist in bookkeeping and the resolution of disputes.

4

The building process in 68 steps

Here it is – the step-by-step guide on how to build a project home. The materials we used were basically a waffle slab, timber frame, clay brick, and a steel roof. Waffle slabs have been used for many years and a large number of project home builders still use them today. However, depending on the size and length of the slab footprint, additional structural steel may be required. Even if you are considering using different materials such as steel frame and roof tiles, it will still give you an idea of the steps involved.

The internet, or Appendix B, will help to find services, materials and tradespeople, and a computer makes it easy to record information on the project.

You've bought your land (Chapter 5), so let's start!

1. Contour levels and geotechnical engineer

Once you have a block of land you will need to have contour levels done by a surveyor to measure the fall of the land. The cut and fill levels will be determined by the contour levels and assist the draftsman and structural engineer to design the slab. A geotechnical engineer will study the soil and perform a borehole test on the land to determine the stability and compaction of the soil and issue a classification. (See Chapter 5.)

The classification of the soil will determine how many piers will be required and the need for structural steel under the slab.

2. Design and cost your house

The working drawings (plans) that you have designed with the draftsman should be completed before moving ahead (Chapter 6). Discuss with your draftsman if a hydraulics engineer is required for your project. Once this is done, contact an estimator to cost your design and project (Chapter 10).

Depending on your council, developer guidelines and land, you may need to order shadow diagrams and landscape plans from your draftsman, but only order these when you have your tender from the estimator and an approved loan so that you know if you can afford to build your chosen design.

It may cost from $10 000 to $15 000 for design plans, land studies, legal requirements, council fees and so on before the slab is poured – basically all the pre-approvals. This is normal, but the estimator may cost only the building of the project. Request an itemised tender of your design and project so you know what has and has not been included.

3. Owner builder course

By law, owner builders need to do a short course to obtain an owner-building permit (Chapter 2). Contact your state authority office for more information.

4. Building permit

You must have a building permit from the state authority office to become an owner builder (Chapter 2). There may be a building surveyor involved to issue a building permit, depending on your state's protocols. A couple doing an owner builder project should obtain the permit in one name only, as permits are issued every five years per person. If you decide to build again within five years you can use the other person's name the second time round, avoiding the need to wait five years before you can get another building permit.

Tip: While you are at your state authority office, purchase some Plain English home building contracts that you can fill out for your subcontractors to sign when they do work on your project.

5. White card course

This course is a short general site safety induction course that is required to be completed by the owner builder. The course can be done online: <www.ownerbuilding.com.au>. It takes about three hours and you will receive a certificate on completion.

6. Finance

If you need to borrow to build your project the best option is to try to get a line-of-credit loan, but only if you can control spending habits and stay within a budget. When you get your tender back from the estimator, you should be able to use the itemised figures for your budget. A construction loan can be time-consuming and inconvenient because the lender will want you to run around and get quotes and write cheques to the service or contractor only. When you have the convenience of a line-of-credit, it is like having money in the bank ready. (See Chapter 3.)

Once you have been approved for the loan and you are happy to spend the money that the estimator has quoted for the project, continue on to the next steps. If not, with your draftsman, modify your design to reduce costs.

7. Engineer for slab construction

The engineer has the responsibility to produce a strong and stable foundation for the project and has to calculate the necessary structural steel and any other requirements for the slab, roof and wall bracing and roof function. Your draftsman will need to have this information to draw structural steel onto the plans. See Chapter 2.

8. Sustainability standards

From May 2011, all new houses in Australia will have to meet certain minimum standards for energy efficiency. NatHERS (National House Energy Rating System) has been set up to assess the energy required for heating and cooling a residential building. The wide range of climates across the country means that standards are set on a state and regional basis; they may also include standards for water efficiency, rainwater tanks, solar panels and planning a home to be positioned in the best location to the sun. See Chapter 11 for details.

In New South Wales, for example, under BASIX (the Building Sustainability Index), all residential buildings are to be designed to use 40% less mains water and emit 40% less greenhouse gases than the state average. To find companies that can issue approval for your design and project, search the internet using the keyword 'BASIX' (or the name of your state's rating scheme).

9. Developer application

Some land developers set building guidelines that people who buy their land are required to follow. The developer application usually requires a set of your plans and the external colour selection you have made. Submit this application before submitting your council application. See Chapter 2.

10. Council approval and colour selections

Every council requires a building or a development application. Again, states vary and may have a protocol for planning permits, building permits and certain professionals who can supply them. The form will itemise the supporting paperwork required for the application. The council will charge a fee based on the cost of your project. They may also require a copy of the tender that your estimator has prepared.

When satisfied, the council will give development approval (DA) with clauses and conditions that you must obey. Each application is judged on its own merits.

While awaiting council approval, start selecting your internal colours. Look at display homes or buy some home decorating magazines.

Council will issue a Commencement of Building Certificate (construction certificate). Inspections are done throughout construction and an Occupancy Certificate is given on completion. There is an option to hire a private certifier to do this work for you, instead of the council. This can save time. See Chapter 2.

11. Housing Industry Association

The Housing Industry Association (HIA) provides a short document on general housing specifications. A private certifier will have a copy of this information. The owner builder and all tradespeople need to follow the document when building.

12. Water authority approval

You will require your water authority to approve your plans (Chapter 2). Some of the application involves:

- service location
- building plan approval
- inspection charge
- building and drainage inspection application.

The water authority may take applications while your plans are still in council. Check with your local water authority. Although this may appear to save time, if the council rejects your plans a new water application will be required for the new plans. It might be safer to wait until council approval is given and then apply for the water authority application.

Tip: This is the time to make contact with the major tradespeople (concreter, drainer plumber, framer, bricklayer, roof tiler) and materials suppliers and tell them about your forthcoming project. Obtain quotes and decide who will be doing the work.

13. Owner builder insurance broker

Contact a broker for your insurance needs in owner-building. A professional owner builder insurance broker will know the policies well. See also Chapter 9.

14. Certificate to commence construction

The DA (Development Approval or building permit or building approval in other states) certificate gives you authority to commence construction. This is to comply with the Environmental Planning and Assessment Regulation. You will need an application form with the appropriate paperwork to get this certificate. (See Chapter 2.) We had a private certifier do this and perform all our council inspections during construction. A private certifier is usually much faster and more convenient than arranging inspections and certificates yourself: see Chapter 17.

15. Retaining walls

Your DA certificate (approved developer application from your council or a private certifier) will state whether you are required to erect any retaining walls prior to construction of the dwelling. Some councils don't mind if you build them after construction. Retaining walls can be expensive, depending on the type of material you use. They are part of the character and style of your home, but it all comes down to money and what you can afford.

16. Site toilet

When you have your certificate to commence construction it's full steam ahead. Now it is time to do the site preparations, such as providing a site toilet. The hire companies will want to sell you a long-term package. Get a few quotes and allow for four to six months construction time for an average-size house. Remember it is not a race because you are the owner builder and are doing the entire project yourself. You don't have all the staff of a building company.

Tip: At this stage you should line up an excavator, drainer plumber and concreter for the slab with an approximate date to start.

17. Peg-out or survey

A peg-out plan is the footprint of the proposed dwelling. It shows the land boundary measurements, north point, and the distance of the dwelling from the land boundary. The surveyor will put pegs into the ground to mark out the proposed dwelling. This helps the excavator and the concreter and locates the correct position of the proposed dwelling.

18. Excavator

Hire an excavator to clear any trees you have been authorised to remove, and to cut and fill your land to a level building platform according to the plans. This will be

the foundation of your slab. There may be some fill left over from the excavation. Ask anyone you know if they need any fill (soil), as transporting and disposing of additional fill can be expensive. The draftsman who does your plans will be able to tell you if this will be necessary. Dumping of soil needs a certificate from the Environmental Protective Agency, who will test the waste for contamination.

If there is rubbish on site it will need to be removed. The excavator usually constructs a sediment control fence as per plans and an area for waste of materials. Discuss this with your excavator.

19. Site signage

You need a clear and visible sign on your site with the following information:

- name of builder
- owner builder permit number
- contact phone number
- lot number.

20. Site access

A temporary blue metal driveway will be required for vehicles driving on and off the land to unload materials. Usually the excavator can provide this.

21. Organising materials and contractors

Now is the time to line up with approximate dates:

- frame
- framer
- windows
- brick supply
- bricklayer
- roof tiles or steel roof
- roofer
- plumber for slab internal drainage.

See Chapters 7, 8, 14 and Appendix B.

22. Piers

Piers are for the foundation of the slab. The concreter will drill the holes in the ground on the land to the measurement the structural engineer prescribed for the piers. The soil classification and type of concrete slab the owner builder intends to use will determine whether piers are required. An inspection is required by the structural engineer before piers are filled with concrete.

23. Forming the slab with concreter and plumber

The footing we used for our homes is called a waffle or a raft concrete slab. This is the most popular slab for project building today. It can be very cost effective and is flexible with ground movement. It reduces the chance of cracking in cornices and brickwork.

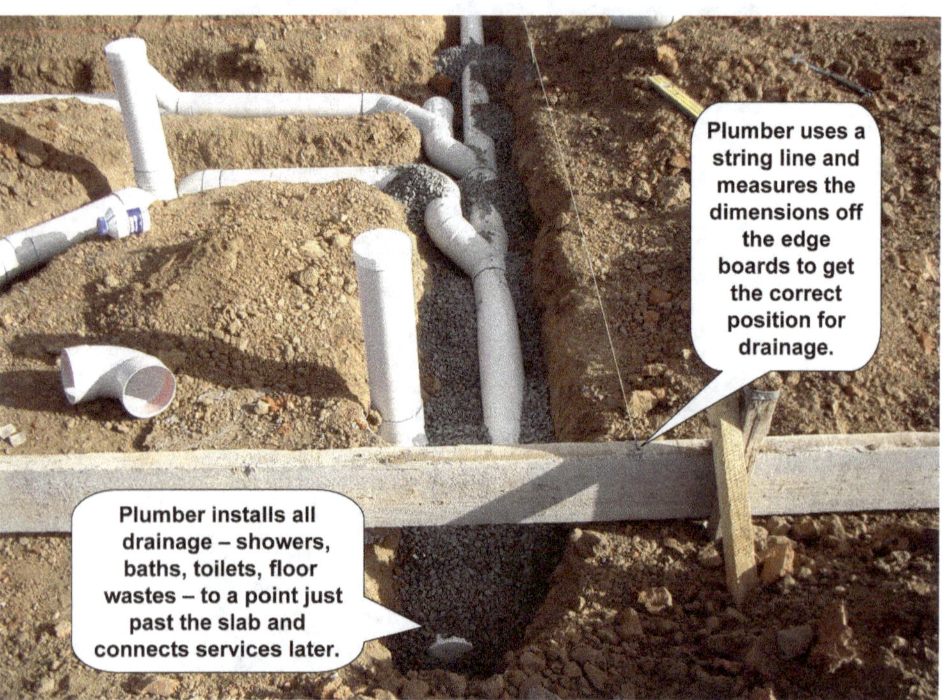

An inspection is required when the edge boards to form the slab are placed, internal drainage rough-in is done and steelwork is in position, and before the framework has been completed and the concrete is poured. The council approval will state when the inspections are required for that particular dwelling. Council can do the inspection or you can hire a private certifier structural engineer with documentary evidence of compliance. The relevant terms of approval and standards of construction (detailed in the Building Code of Australia) need to be obtained before progressing to the next stage.

Tip: If you want to recess your power box you must install a conduit below the ground and between the edge boards before the slab is poured.

24. Pest control

The type of protective termite system chosen determines what happens next. Some systems have a tube covered with a cloth around the boundaries of the slab, and are connected to a box on the ground. This is called Termimesh. Another system, called Graniguard, has a gutter of granite granules next to the brickwork around the boundaries of the dwelling. Discuss the benefits, requirements and council inspections for pest control with the pest control company. Pest control collars or termite flanges may have to be fitted on the internal drainage around this stage. It is important to do your homework first and check if the proposed site is termite prone. If so, further works such as barriers and chemicals may be required.

25. Pour the slab

After inspection of the piers and the slab preparation from the structural engineer, the concreter can pour the slab. Ask the concreter to apply a curing compound on the slab to reduce the risk of cracking in the concrete. This holds the water level in the slab and stops rapid moisture loss.

Concrete being poured and
pumped with a line pump.

The day after slab is poured
apply a curing compound. It
will stop the new slab from
cracking. Acrylic based
compound is the cheapest
and will do a great job.
Other methods of curing
can be done. Check with
the concreters prior to
pouring the slab.

26. Excavation and connection of services

Contact a drainer plumber for the excavation of trenches and connection of sewage, water and stormwater drainage. This will connect the dwelling to the sewer, water and stormwater junction points near your property. Ask the plumber about installing an inch supply pipe to improve the mains pressure to the house. **A council inspection is required before covering trenches.**

Tip: Take a lot of photos for your records in case you need to locate the pipes in future.

Storm water
Sewer line

27. Temporary fencing

Safety regulations require a temporary fence around the property boundary. Again, you will probably need to take out a package to get a good deal. From this stage you should consider hiring the site toilet for a minimum of four to five months of building time.

We used reinforcement mesh SL72 to create a temporary fence. After we used the mesh for this purpose, we used it again for the driveway reinforcement and retaining walls.

28. Erect the frame

It is time for the frame to go up. It is an exciting part of the building process to see your design come to life with the shape of your lounge room, bathrooms, bedrooms and so on. The framer will fit the windows, sliding doors, external door architraves and roof trusses.

Contact a company for a roof safety rail now.

Tip: Buy a cheap toilet (you should be able to find one for about $100.00) and install it temporarily without a cistern on your connected sewer pipe. Then put plastic around it for privacy. Use a bucket to flush. That way you don't need to hire a portable toilet system.

OWNER BUILDER
Sediment control fence installed along with an on-site toilet and owner builder sign with licence number and contact number.

Windows have been installed by the frame carpenter. Tape up the window frame edge so the bricklayer doesn't get mortor on it (it's hard to clean off).

Frame carpenter installs
sliding doors.

29. Electricity

Connect the electricity service box to the frame. An idea we had with this electricity box was to build it into the frame and brick around it, so that the box will only be slightly forward of the completed brick work.

30. Plumber and gas rough-in

Call the plumber to return for the plumbing rough-in. This is the placing of pipes for water and gas throughout the house before the plasterboard is placed against the timber frame.

The kitchen and bathroom pipes need to be fitted before proceeding further.

Tips:

- If the plumber does not use silicone in every drill hole through the frame, do it yourself to ensure that you don't have clunking pipes. Pipes can also be secured with fixed bracing and lagging.
- After the plumber has completed this procedure we had a special type of external pink insulation (Gladiator Tough Wall Wrap) wrapped around the frame. It helps to reduce noise, dirt and spiders, and helps with heating and cooling the dwelling.

If you plan to have a gas hot water system, attach the recess box to the frame before the bricklayer starts.

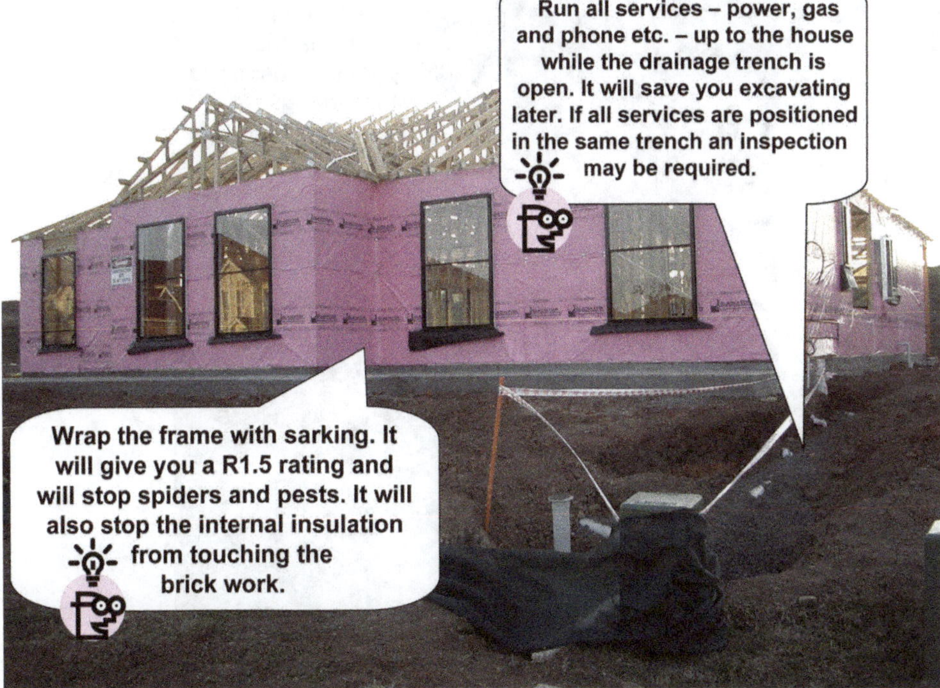
Run all services – power, gas and phone etc. – up to the house while the drainage trench is open. It will save you excavating later. If all services are positioned in the same trench an inspection may be required.
Wrap the frame with sarking. It will give you a R1.5 rating and will stop spiders and pests. It will also stop the internal insulation from touching the brick work.

31. Water tank

If you are installing an in-ground water tank, it would be ideal to do it around this time. If you are installing an above-ground water tank it can wait until near completion of the project.

32. Bricklayer

The brick supply should now be on site – the bricklayers' sand, cement and hardware. The bricklayers' hardware includes:

- lintels (structural steel)
- brick ties
- bycol (an air entraining admixture), and oxide to colour the mortar.

If you are building a two-storey or higher dwelling, you will need to provide brick layer scaffolding.

Now we need the brick layer.

Brickwork started – you will have to arrange lintels, damp course, brick ties, Bycol (used to improve the workability of mortar and concrete) and oxide if you want to have coloured mortar.

OWNER BUILDER
The brick work is beginning to take shape.
WARNING

Lintel placed prior to the
brick gable being
Installed.

View looking
inside the
brick cavity.
Make sure the
bricklayer
washes out the
cavity every day
so the mortar
that falls
down does not
build up.
Be careful not to
block the weep
holes or window
flashings

Tip: Our bricklayers received a case of beer from us every Friday afternoon and a cheque for the hard work they had done all week. The brickies returned the favour to us in their workmanship on the dwelling.

33. Metal fascia and gutter and roof installation

A safety rail is required for the roofer to fit the metal fascia, metal gutter, roof sarking and battens, roof tiles or steel roof. In really hot weather the roofer may not be able to work, especially with a steel roof. The metal gets too hot to handle.

Steel roof
progressing well.

DANGER
Steel roof complete.
Now is the time to order the
ceiling insulation.

34. Planing the frame and nail-off brick ties

The framer will adjust and plane the frame to ensure the plasterboard (internal walls) will lie smooth. At this stage the brick ties need to be nailed off and secured.

35. Council inspection

An inspection is required when the framework including wind bracing and roof members is in place, and immediately after the frame has been completed and before any internal walls are fixed. Remember to check council approvals for scheduled inspections.

36. Clean site

Clean up and keep the site tidy and hazard-free. A lot of off-cuts of bricks and timber will pile up. We did our own site cleaning.

37. Eave carpenter

Eaves, pergolas and hardware materials should be prepared for the eave carpenter. As we built a bullnose style roof, the eave carpenter's needs were minimal.

Eaves installed

38. Electrical rough-in

At this stage, all the ducted air, ducted vacuum, alarm, communication cables, speakers, home networking and phone can be prepared.

Tip: Run a category 5 networking cable (cat 5 cable) down the cavity with every phone point and antenna point. Leave approximately 20 m rolled up in the roof. This will come in handy later if you want any home networking equipment. Take photos to record where the cables and wires run.

39. Brick cleaner

Now the brick cleaner can do his job. If you decide on face brick (no coating) the bricks will need to be acid washed to clean the over flick of mortar. Sometimes a mild acid is used to clean the bricks. Discuss this with the cleaner, especially if you are going to coat the bricks with a render, because this is a specialised area that needs to be discussed with the company or tradesman who will be doing the work. Render is very expensive compared to bagging cement and it would be best to leave it up to the company that takes on the job because they would need to give a warranty. We did not need a brick cleaner because we had bagging to the brickwork.

40. Wall insulation

Dress up in an all-in-one suit if you are going to do this yourself. You will be itching and scratching for days if you don't. If you are a light sleeper it is a good idea to insulate around your bedroom walls to help suppress noise.

Tip: In every corner of the house where there is a hip or trusses, install ceiling insulation and suspend it with a string as it is difficult to install the insulation in the corners.

41. Plasterboard internal linings

The best place to store your internal linings out of the weather is in the garage - hopefully close to the house.

Plasterboard being installed.

42. Metal downpipes

Only install metal downpipes at this stage if you are having an external brick face. If you are coating the bricks, that is having rendering applied, install the downpipes later. Otherwise the brick coating will be all over the downpipes.

Brickwork bagging completed.

Bagging complete – just have to paint now.

43. Staircase and garage door

Install the garage door to keep your house secure. Install the staircase if you are building a two-storey or higher home. Tell the stair company that you don't want the staircase to creak as there are extra measures they can take to avoid this noise. As soon as it is installed, protect the staircase with a temporary cover.

44. Install hobs for shower and bath

Depending on the type of shower and bath you have chosen, the floor tiler should be able to do this.

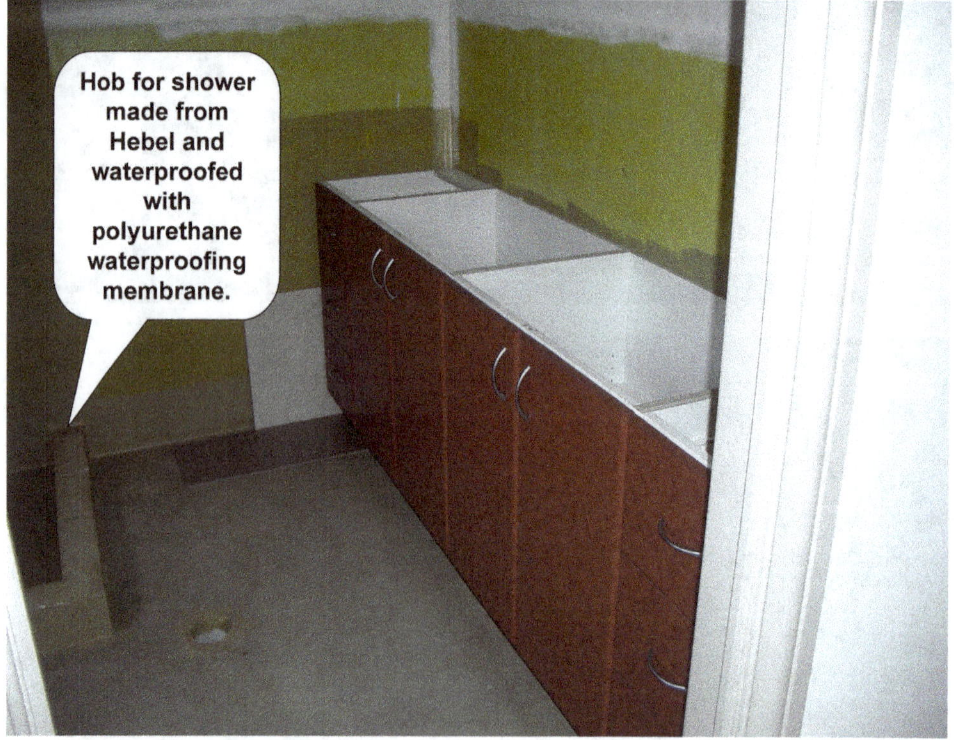

45. Waterproof flashing in wet areas

The flashing is a special coating and material painted on to waterproof the wet areas of your home, as required by law. Polyurethane membranes are the best, but check that the tile glue will stick to the membrane before it is installed. The owner builder can attend to this job, but a professional is highly recommended.

Waterproofing membrane installed. Make sure you check the compatibility of the membrane with the tile adhesive.

Hob and step for bath made from Hebel (aerated concrete) with CFC (Compressed Fibre Cement) sheet on the top. Simply glue on tiles directly.

46. Council inspection

This inspection is required before tiling floors and walls of wet areas.

47. Final fix-out – carpenter

The carpenter fits items such as wardrobes, doors, jambs, linen cupboards, pantry, door handles, hanging rails, towel rails, toilet holders, skirting and architraves.

48. Install bath

It is a good idea to install the bath now while there is room to manoeuvre, so it will not be a struggle when the shower, toilet and vanity items are in the bathroom.

49. Tile wet areas

It is best to tile now, as the toilets, vanities and showerscreens will sit on the tiles.

50. Install toilets and shower screens

If all plumbing is complete in the toilets, you can start using them now and remove your on-site toilet. The supplier will fit the shower screens in position and do a fine job. The shower screens need to be waterproof and must be made of safety glass. When the shower screens are fitted, the company needs to provide a safety glass shower screen certificate.

Tip: If you want to avoid water dripping onto the floor tiles when you open your shower screen door, consider having a sliding door rather than a pivot door.

51. Install kitchen, vanities and sinks

The kitchen and vanities are made off site. It is only a matter of having them delivered and fitted.

52. Plumber finish-off

The plumber can now fit all the tapware, shower sets, bath sets, toilet systems, dishwasher and ice maker for the refrigerator, connect an above-ground rainwater tank and install external taps for the gardens. Ask the plumber if he can install the hot water system now.

53. Painting

The final result of the painting in your home will really depend on the technique and skill of the painter. If you find a painter by word of mouth you have a better chance of finding someone with a good reputation. A paint job can be either really good or not so good. Builders have trouble trying to find a good painter. The old-fashioned way of painting is best – with a brush and roller. If you are still very motivated, you can give it a go yourself. We painted our house to save $4000 but it is a lot of work and time consuming.

54. Electrical fix-out

This is a big job to be done now by the electrician – installing ducted air vents, alarm sensors, lights, switches, smoke detectors, exhaust fans, TV points, wall oven, range hood and powerpoints, and connecting everything to the electrical board. A smoke detector certificate is required from the electrician.

55. Floor – sand slab

This helps the floor coverings to sit level and is especially necessary for vinyl floor coverings, timber flooring and tiles.

56. Clean site

It is a good idea to have a mini clean up, internal and external, to remove all the off-cuts from the tiling and carpenter fix-out. We cleaned up the site ourselves.

57. Floor coverings

Lay the floor coverings after you have performed the mini clean up, as until then there will be a lot of dust around and you don't want floor coverings such as carpet to be soiled.

Finished and just needs a good clean.

All that needs to be done now is to install the bullnose verandah.

58. Final pest control

Discuss the final stages of pest control with your pest controller. We had our house sprayed by the pest controller at this stage. There were a lot of spiders around because of the disruption to the ground on-site. A certificate must be supplied by the pest controller stating that termite protection has been installed and there is no evidence of infestation.

59. Driveway and paths

Before pouring concrete, obtain approval from the council for the crossover from the road and over the nature strip to the beginning of driveway.

Tip: Dig a trench below the driveway before the concrete is poured, so you can run irrigation and lighting afterwards.

60. Roof insulation

Wear protective gear if you are installing the ceiling insulation, and keep the insulation away from any hot transformers or down-lights in the ceiling. Protective boxes and LED-type lights are other options to avoiding a fire from heat in the ceiling.

Only a licensed roofer can apply the roof sarking insulation blanket.

61. Mirrors

Install mirrors to ensuite and bathrooms after wall tiling or paint work have been done.

62. Final council inspection

When the dwelling has been completed the final inspection must be done before the building can be occupied. Once approved, an occupancy certificate will be issued. Council or a private certifier can do this inspection.

63. Professional clean of house

You can do this yourself or hire someone. It is a big job to do and yes, we did it ourselves.

Clean:

- bathroom tiled areas and vanities
- kitchen cupboards
- windows, light fittings, mirrors
- floors, skirting and architraves.

64. Kitchen appliances and hot water system

If your house is secure and has an alarm, install the kitchen appliances. If you have not already installed the hot water system, do it now.

Tip: Check all under-sink connections are tightly sealed. When we installed our hot water system we had bit of a flood in the house because the connections were not sealed properly.

65. Blinds and light fittings

It can take up to a few weeks before delivery and installation of blinds. Discuss this with the company to find out how long this might take. Light fittings are easy to install if you are not buying fancy ones.

66. Fencing and retaining walls

The retaining walls will need to be done first, of course. The fencing cost is to be shared between you and your neighbour. Usually three quotes are done and then a decision is made by both parties.

67. External accessories

Install the mailbox, clothes line, security doors and fly screens.

68. Laying turf

Don't lay the turf until you move into the property, because you need to be there if you are to water it every day. Your local water authority will need to provide you with a water approval if you are in an area that has water restrictions. Prior to laying turf, check that the site drainage is positioned on the uphill side of the building. This will avoid water pooling next to the house and under the slab. The geotechnical consultant should have considered the roof drainage as per the Building Code of Australia.

5
Choosing your land

Location, location

Should you buy the land first, or should you choose the house first and buy the land later? Either way you must consider the location: where do you want to live? Of course this will depend on a number of lifestyle choices – if you have everything you need where you are right now, it will be difficult to leave the area with all the familiar amenities, but if there are no vacant blocks of land you will need to consider moving somewhere new.

Think about what you need for you and your family such as:

- Do you need land somewhere close to work?
- Do you need it to be close to your children's school – are there primary or high schools in the area?
- Do you want to be close to a hospital, medical centres and doctors for children and elderly parents?
- Is there public transport? What transport is available in the area?
- Do you need to be close to public transport? And how long will it take you to get where you need to be?
- Do you need access to a child-care centre?

You will want to be in an area that has local shopping centres. Other desirables may be bigger department stores, sporting clubs, and recreational centres such as parks, tennis courts and swimming pools for you and the children.

One of the most popular places to find vacant land is in a housing estate. These estates create whole new communities and they build schools and shopping centres close by. Many have loads of other amenities like swimming pools, tennis courts, basketball courts and walking tracks; some also have club houses. They usually have associated fees for the upkeep and maintenance of a swimming pool, so remember to ask what you get and how much it is. Some estates have shops and restaurants. If you are looking for a large block a golf course estate may suit you, as the blocks of land are usually much bigger than in the average housing estates.

These are some of the many questions you will want to consider. Write out a list of what you need and what you are willing to compromise on.

Constraints and restrictions

Another really important question is whether you want a single level or two-storey house. Some blocks of land are too small to build a single level home. If you have your heart set on a particular home or you must have a single level, when you find a block you should take the block plan to the builder to draw it up in accordance with your council or estate setback requirements and floor space ratio. In some new housing estates where the blocks are only as big as 330 m^2 and therefore only two-storey homes are allowed.

Your next consideration is whether there are any restrictions on the land. Any restrictions can cause you problems down the track.

Before you sign on the dotted line it is best to obtain a title search. This document will will tell you whether the seller has properly disclosed any 'notifications' on the title like an easement or any restrictions to the use of the land, and it allows you to be sure the details in the contract are accurate.

The title will give you such information as:

- the names of the owner(s) of the land
- description of the land
- easements
- covenants
- mortgagees.

The owners

The first piece of information is pretty straightforward: it states who the current owner is and who the previous owner was if it is land that has recently been subdivided.

Description of the land

The description of the land is also straightforward: it has the address and the plan number, which you need to know to do a title search. The plan number has different names in different states so it might be called the deposited plan (DP), registered plan, linen plan or certified plan. The title will tell you the zoning: this refers to what you are allowed to build on the block. If you want to build a duplex and you are not zoned for it, you will not be allowed to build one. There are a few different types of zones, like residential, commercial and rural. For each of these zones different rules apply, so if you have any questions about what you are allowed to build on your block, visit your local city council.

Easements

An easement is something that restricts your ability to use the land. You need to know about easements because you are not allowed to prevent access to them. There are some exceptions to this rule though; you may be allowed to erect a shed or some other type of structure as long as it does not interfere with the ability to service the easement. Types of easements include:

- a pad-mount, which is an electrical inspection box
- a drainage easement
- sewer or service easements
- a right-of-way.

An easement for services allows the defined strip to be used to carry services such as electricity, gas, water and telephone, or sewerage on that part of the land. In new estates easements carrying services to neighbouring properties are mostly at the front of the block, but it is important to note where they are.

Rights-of-way can be a right of carriageway, which allows vehicles to use it, or a right of footway. The most common situation in which a right-of-way is found is with battle-axe subdivisions, where one or more blocks have the right to use a defined strip of land for access, such as a driveway to your lot. The dimensions, position and terms of the easement to right-of-way are usually defined when a subdivision is registered at the Land Titles Office.

It's important to find out about easements before you buy land because not only can they affect the value of the land, depending on the nature of the easement they could cause you some inconvenience. If you plan, for example, to build a swimming pool on that part of the land that would be illegal and the pool might be removed at your expense. However, in most cases you should not cancel your purchase based on an easement – better to find out if it is going to really be a problem.

Covenants

Covenants are usually limiting too, because they can stop the owner or prospective owner of the property from doing something. For example, a covenant can specify things like:

- minimum floor size of your home: it may control what size home you can or can't build; usually the minimum size is something like 'no less than 24 sq'

- the types of materials that must be used, such as brick construction; some guidelines recommend you have some type of light-weight cladding in the upper level.

Some covenants protect your investment because they regulate the standard of construction allowed. This can be useful if your neighbours decide to paint their house yellow with pink polkadots.

You need to understand what the covenants mean to you and your block. It is advisable to seek advice from a qualified conveyance or solicitor if you are not sure.

Mortgagees

The title also lists the person or institution that has a caveat or hold on the block if you have borrowed money from the bank.

Other restrictions

It is really important not to get caught up in the excitement of purchasing your land. There are many drawbacks if you are not careful, and these may cost you many thousands of dollars. The following lists some more restrictions that might be associated with your block.

- Local government policies in addition to standard requirements may reclaim your land in the future, for instance to build a road or airport.
- Flood plain issues: are you in a 1:100 flood zone or any type of flood area? If so you will have to build your home at a higher level above the flood plain.
- Bushfire considerations: if you are in a fire-prone area do you need to incorporate certain building materials in the construction of your home?
- Is the site in a seismic area requiring specialised engineering requirements – are you on a fault belt that is known for earthquakes?
- Wind rating: there are several rating classes that may require you to take into consideration window choices, doors, roof construction and so on.
- Trees in the building envelope and gaining approval to remove them, plus the cost.
- Is there soil salinity in your area? If so you will need to address this by providing a protective damp-proofing membrane.
- If are you thinking of building on a battle-axe block, are the services like water, electricity, sewerage, gas, ready on your block, or do you have to pay for them to be brought up to the block?
- Does your block have rock or sandstone? Drainage is also very important: you may have to dig trenches and put in rubble or create an easement of your own to divert water from getting into the house or undermining your foundations.
- If you do not have town water and sewerage, you will have to provide some type of system for refuse treatment, e.g. septic tank or leach drains.

Consideration should be given to whether they will pass health tests. You can investigate the different types of systems available through the internet or by approaching your local council. If you don't have town water a rainwater tank may be required. They come in many sizes: if you plan to build in a rural area you may have to have a really large tank, one for your household needs and one for firefighters.

- What does your block look like: is it flat, does it have a bit of slope, does it slope to the street or does it slope to the back? How much cut and fill is allowed within your council guidelines? If you purchase a block of land because it is cheap, you may find out later that the extra cost to build your home will be more than you saved by buying a cheap block. If you are building on a steeply sloping block you may run into all sorts of problems, like having to construct a split-level home, sometimes with as many as 10 steps internally: this also means you must increase the ceiling height to accommodate the steps. Then there are retaining walls on top of that, and your slab will have to include drop-edge beams.

Every one of these conditions can significantly impact on the cost to build your home.

The list of easements and covenants we have provided is in no way exhaustive, so you simply must do your research on your prospective block. It is best to seek advice from a qualified conveyance or solicitor before the land contract is signed.

The importance of homework

One of my clients, Debbie, had her heart set on building one of my project home designs in the Blue Mountains, New South Wales. Her husband was a musician, and they wanted to be able to play without having to worry about neighbours complaining about the noise.

After looking at several blocks over a period of months and sending me contracts to look over, Debbie was really excited because she had finally found one that was the right price and had a slab already to go. We wouldn't use the slab that was there, but she could build a studio there later once her house was finished. She was prepared for a septic system and 15 000 L rainwater tanks and all the other restrictions involved in building in rural areas except for one thing – the distance of the house site from the road (about 30 m) meant that she would be required to build a proper road with an asphalt base, presumably to provide access for the Rural Bushfire trucks in case of bushfire attack. Unfortunately this would cost her another $30 000, which was not in her budget. So we had to cancel her tender request.

Debbie was sensible, and had done her homework, so she hadn't exchanged contracts and wasn't out of pocket.

A cautionary tale

Another of my clients had his heart set on building a dual occupancy. He had purchased a block with an existing home at the front and wanted to build a home behind this one (a battle-axe site). The land he bought was 16 × 45 m, so by today's standards it was pretty large. He wanted to retain the home at the front as he had tenants living there and they were paying his mortgage, and he couldn't afford to have that house vacant.

There were a number of problems with his development because he had not done his homework.

1. When you want to build on a battle-axe block you need a 3.5 m wide driveway for a car to drive to the second home. Unfortunately the house at the front only left 3 m so we would have to cut off 500 mm off the house – which we couldn't do.
2. Even if we had a 3.5 m wide road to get to the second home, we had to provide a turning bay for the car to reverse out of the garage and drive out in a forward direction, as you are not allowed to reverse 30–45 m. The existing house was too far back to allow us to build even a really small home once allowance had been made for the 6 m required for the turning bay and the mandatory 80 m^2 allowance for private open space.
3. The existing home also had to have its private open space, and as we had to take 3.5 m from the side for the driveway (this cannot be included as private open space) we could not make it without taking some space off the one at the back – which made the situation for the prospective house at the back even worse.
4. The other problem was even if everything did work out and we could develop the block at the back, my client wanted his tenants to remain renting the house at the front. This wouldn't be allowed: the services would have to be closed off until everything was ready to connect for the house at the back, and this would take a few weeks.

My client had thought that because the block was the right size for dual occupancy and he had a reasonable frontage that he was on the right track. There wasn't anything I could do to make it work for him. I spent two or three hours with different designs but I couldn't help him unless he demolished the home at the front and built two new homes, and he could not afford to do this.

I was really upset that he had wasted his money. If he had gone to council before he bought the block and had done the right research, he would have realised it couldn't be done. He did buy the dual occupancy development control guidelines, but he just did not understand everything that was involved. On top of this he was faced withcapital gains tax if he sold.

Soils and slabs

The next consideration dictated by the site is what type of slab you need. You also need to think about where to orientate your home, as if the siting is not favourable there will be extra costs associated with the choice of land.

The slab type depends on the type of soil you have. Your house designer will have an engineer who will determine the strength of the slab required and type, taking into account the type of soil, the load and the slope of the land. Some slabs are better suited to different soil types (see Table 5.1).

As well as soil type, the slab you choose will depend on the slope of your block and the load of the house design, especially if you want a double brick home.

Your slab is an integral part of the foundation for your home. That's why as well as the soil classification you need a contour survey to determine the slope to ascertain the best slab for your home and site conditions. The type of slab used may also depend on the cut and fill allowance determined by your council. Concrete slab construction centres on footings poured for support. Reinforcing materials set within the concrete help prevent cracks and settling. Australian Standards 2870 slab (this refers to the thickness of the slab) has been approved to be adopted in 2012 and applies to all new constructions.[1]

This new standard places particular emphasis on design for reactive clay sites that are susceptible to significant ground movement due to moisture changes.

These standards take into account the following:

1. Swelling and shrinkage movements of reactive clay soils due to moisture changes;
2. Settlement of compressed soils or fills;
3. Distribution to the foundation of the applied loads; and
4. Tolerance of the superstructure to movement.[3]

What this means is that if you have a geotechnical report (not a borehole sample, which is not as accurate) that identifies your land as an H type soil (highly reactive, meaning it will swell due to moisture and crack due to high temperatures)

Table 5.1 Soil classifications[2]

Soil class	Description
A	Mostly made up of sand and rock and has very little change when wet
M	Moderately reactive clay with moderate movement due to moisture changes
S	Slightly reactive clay in the soil with only slight movement due to moisture changes
E	Extremely reactive site with extreme movement due to moisture changes
H	Highly reactive clay site with high ground movement
P	Partially reactive: soft soils, land slip, mine subsidence collapsing soils due to moisture and soil structure changes

you require an H type slab. The H type slab has now been split into two classes, H1 and H2; under AS2870-2011 you must provide an N16 slab (using thicker bars) for an H2 classification more than 20 m long instead of the N12 slab specified under the old AS2807. Even if you are over the 20 m mark by only a few centimetres you cannot simply put in an expansion joint at the 20 m mark and build the remainder as N16, the whole slab must be N16.

The reason behind this new thicker slab is that a number of surveys and tests have been conducted on Australian housing footings and foundations and shown that the reactivity (shrinking and swelling) of the clay soil can be detrimental to footings or slabs. The slab needs to be designed to counteract the movements of the clay soils, otherwise damage such as cracking can leave the slab open to pests such as termites and require expensive repairs. It is better to invest in a better, stronger and more flexible slab at the construction stage of the house.[4]

Since we are not experts on foundations we have listed the references and websites we used for our research to help you when dealing with this aspect of being an owner builder. The most common options for slab construction are described below. There is also some information on other slab types that suit extreme sloping blocks and difficult soil types. Construction materials and methods for slabs are described further in Chapter 7.

Stiffened raft

The stiffened raft slab is the simplest and most common slab construction available. The stiffened raft configuration can be used on all classes of sites except problem sites (Class P).

Stiffened raft slabs consist of:

- 100 mm thick concrete slab
- edge beams
- internal beams (except Class A and Class S sites)
- steel reinforcement.[5]

The stiffened raft is a slab that is poured onto the ground with concrete edge and internal beams. The whole slab is poured at the same time.

Waffle raft

Waffle raft slab construction has slab ribs formed on top of the ground using a grid of polystyrene void forms are laid out on a levelled area. It is suited to reactive clay sites.

If the weather has been really hot and dry over period of time reactive clay soils tend to compact and shrink, resulting in cracks in the ground. In our back yard we have had cracks big enough to put a hand in up to the palm, which would be at least 5 cm deep. When we have a lot of rain this causes the soil to swell, and hence we need to use a slab option that will move with the weather conditions.

Pier and slab

Pier and slab is a slab system that uses a pier grid into the soil and then the slab poured on top. The holes are drilled down into the soil to the engineer's specifications and then filled with concrete. Then the slab is poured on top of the piers, connecting the piers and the slab. The size of the piers and the depth depends on the soil type and the load of the house; these will be detailed by a structural engineer on the day of the pour. This is the most common slab that builders use.[6]

Strip footings

Concrete slabs may be supported by strip footings (also called edge beams). A footing slab requires two separate concrete pours. It can only be used on Class A and Class S sites.

Advantages of strip footing slabs are that they:

- adapt to sloping sites
- require simple formwork (no edge rebates)
- require simple excavation that is exposed for minimal time
- do not require internal beams.[7]

Strip footings are a reinforced concrete beam installed into soil and laid around the entire outside of the house. They are the foundations required when building a double brick or brick veneer house with a timber floor. The outside walls of the house are bricked up to floor height, and then the floor is installed. For this to be able to be done correctly there must be a strong foundation for the house to sit on.[8]

Useful websites

www.concrete.net.au
www.ehow.com/list_6712497_types-slabs-construction.html#ixzz12DMPQO2V
www.lysaught.com
www.meshbar.com.au
www.onesteel.com
www.sustainability.vic.gov.au
www.thinkbrick.com.au
www.yourhome.gov.au

6

House design

Wants and needs

Homes embody much more than just four walls, a roof and a few bedrooms. They provide us with a place to relax after a hard day, and they allow us to socialise with family and friends in our own space. A very popular feature incorporated in many current designs for homes is an alfresco living area that opens directly from family rooms, bringing the outdoors in and giving us lifestyle choices.

Building a home is one of the biggest financial decisions the average Australian will ever make. You will no doubt want to create your home with everything you need. This can be an overwhelming task. The choices are enormous so the best way to figure it out is a wish list (not a bucket list). Keep in mind you will want to stay within a budget that you can afford. Start by writing out a list of must haves, maybes and your wish list, in three columns. Table 6.1 shows an example.

We all want a lovely big house with big spacious rooms, but do we really need it? How much time do we spend in a bedroom, for instance? Instead of spending money on a bigger bedroom, wouldn't it better to spend it on making a family

Table 6.1 A sample wish list

Must have	Maybe	Wish list
4 bedrooms	mud room	billiard room
master bed en-suite	laundry chute	prayer room
walk-in pantry	shower niche	balcony

room bigger for everyone to use? However, this may be something you have always dreamed of, so if this is a priority try to use good design and use of space to include everything you absolutely need. There isn't any point in designing a big home if you won't use it, and bigger isn't necessarily better. These days the formal living and dining are not so popular because people realised that they did not use them – they were just two more rooms to furnish, clean, heat and cool. You may find you will gain much more spending that money on an alfresco for outdoor entertainment and/or a fantastic garden or swimming pool.

When you have a list, go over it until you have refined it to what you really need and what you can afford. Also bear in mind that trends change and what is in fashion now could be superseded in a few years, so stick with safe and affordable choices. Some of our clients would ask for a price in a tender for everything they want or need so they have an indicative cost, and then take out what they can't easily afford.

It's a good idea to visit new display villages to see the latest trends. Pick up the builders' brochures. They will tell you how many square metres the house is, and most have the dimensions of the rooms written on them. This will give you a good feel for room sizes and the way furniture can be used. Put this all together and you can judge for yourself if it is the right size for you. This will give you a good perspective on the base price too (keep in mind that what you see in display homes is not what you get. They are all loaded up with the best of inclusions. Don't forget they add the cost of whatever promotion they have going to the base price, and remember that the builder usually works on a 20–30% margin.

Exterior design

Find out from council the requirements for homes in your area. If you are building in an estate they often have design control guidelines as well. One of the requirements will be size; if there is a floor space ratio that may restrict the size of the home you might want to build. These controls relate to bulk and scale, and are a means to stop your home impacting on other homes on either side of you. Additionally they control the types of building materials, façade style, fences,and other external features of your proposed home. (See also Chapter 7.) These restrictions will be clearly stated in the control guidelines so it is important to study it and the implications to your development. In any of the newer estates, if you had your heart set on a style vastly different from the surounding houses it would most likely not be approved.

Many people choose to add a balcony and or verandah to the façade of their home. This provides relief from long straight masonry walls (articulation) and improves street presentation. Corner lots have special requirements to break up long masonry walls, and providing articulated relief and changes the roof line with a gable roof or by adding a wraparound verandah, bay windows, a corner entry, awnings or pergolas will give your home presence from the side street.

Another way you can make your home stand out is the choice of roof design, with options such as a pitched roof with gables, hips or a combination of both, flat roofs and curved roofs. For energy efficiency use light-coloured roof materials: dark colours absorb more energy and will store unwanted energy in the roof cavity.

Your draftsman or architect will be able to help ensure that your house design and positioning comply with all of the relevant building regulations in your area and meet the criteria of any design guidelines or covenants set by the developer.

Make sure you have the correct width and depth for your block and know your side back and front setbacks. Corner allotments will have a secondary setback. If you have an easement on the block, find out how close are you allowed to build to it. Many easements require a certain distance to be free for maintenance access.

There are loads of websites for you obtain information to design your home to make best use of your block. Start with <www.yourhome.gov.au>: this website is invaluable, with guidelines and illustrations to follow on how to make the most of your design within the confines of any block.

Solar access

The size and shape of the land can be an advantage or disadvantage to solar design principles. Ideally the land should enable enough flexibility to orientate a new home east–west to increase the northern aspect.

A house designed with the effects of the sun in mind captures the radiant heat of the sun during the winter while avoiding excessive heat build-up in the summer. Solar design also considers the sun as a source of home lighting. Planning with this awareness results in a home that is naturally more comfortable, and once you learn the principles of an efficient home the rest will flow from there.[1]

The best orientation will enable you to be more energy efficient by using these principles to utilise the sun and shady spots to your advantage. Passive design uses the natural climate to its fullest; without the need for additional heating or cooling input or at the very least reducing your need to use artificial heating and cooling. This makes passive-designed homes more comfortable to live in, year round, and results in reduced energy use and therefore reduced costs. By designing your home to be as energy efficient as possible you may save many thousands of dollars over the life of your home.

If you have a north–south facing block, it gives you good access to the northern sun with minimal overshadowing from your neighbours' homes. Eaves are especially important in planning your home: they shade the windows in summer and in winter they will allow the sun to enter. It is best to orientate living areas to the northern side of your home, thereby having lovely sunny rooms in winter; and in the hot summer months, when the sun is high in the sky, the use of eaves, pergolas and verandahs protects this aspect from the sun entering your home.

The floor plan

If you have been to display villages and looked at builders' magazines you will have an idea of what you would like included in your home besides the obvious living area, kitchen, bathroom and bedrooms. You will have a distinct concept of where you will want these rooms to be laid out for ease of use and outdoor living. If you are planning a family you may want a small room close to your room for when the baby arrives; if you have children you may want them as far away from your bedroom as possible. Shift workers unquestionably want their bedroom at the back of the house so they don't have to hear cars driving up and down the street. If you plan on staying in this home for a long time, plan for your family's needs in the future: do you require bigger bedrooms for the kids so they can study in their room, or do you want a study nook where you can watch what they are doing while you cook? As they grow up, you may want them to have a dedicated room to play games away from your quiet space.

The best placement of rooms and construction materials will go a long way towards attaining comfortable use of rooms. It is important to prioritise rooms based on solar penetration. An open-plan kitchen and living area, for example, should have the best situation for lighting so you don't have to have lights on during the day, while bedrooms or bathrooms need less daylight as they are mostly used for small amounts of time, or at night.[2]

By zoning your home and providing doors you are enabled to close off unused rooms. This will save on extra cooling and heating appliances.

Kitchen

The centre of your home is unmistakably the kitchen. This is where your family will spend a great deal of their time. You may also want to include a living area and an alfresco area off the kitchen so that when you are socialising with friends and family you will still be part of the festivities while you are stuck in the kitchen whipping up the evening meals.

A good functional design with plenty of bench space for food preparation is a must. It is always handy to have a bench top with work space close to the fridge and pantry. It is also a good idea to locate the sink and dishwasher in the same bench as this saves money on plumbing. If you plan on a refrigerator with an ice-making facility, remember to have the plumbing installed at the same time. Make sure kitchen cabinets allow a decent air gap around the fridge (especially at the back) as it needs good ventilation to work efficiently.[3]

Bathrooms

There so much to love about bathrooms – the vanities, taps, tiles, baths, showers, towels, tops and mirrors. There is so much choice. A good place to start is by visiting showrooms and display homes to get ideas of the latest trends and designs.

Showrooms and display homes will also have the latest in water-saving taps, showers and toilet options to help reduce the use of potable water. Showerheads must be a minimum AAA rating. An old-style single-flush toilet could use up to 12 litres of water per flush, while a standard dual flush toilet uses just a quarter of this on a half-flush. As technology advanced we now have brought that down to 4.5/3 L flush, achieving a WELS rating of 4 and 5 stars in Australia. This equates to saving the average household about 1000–1500 L of water each year.

Careful choices in your bathrooms and laundry will add value to your home. With all the latest innovations you can save lots of water in bathrooms and laundries without having to compromise a thing. Your plumbing fixtures will last a long time, so it pays to get the best in water savings. Having windows that open to ventilate bathrooms and laundries is cheaper and quieter than relying on an exhaust fan.[4]

A rainwater tank can save on water use if it is fitted to flush toilets in lieu of using mains water. A grey-water system can capture bath water/ shower water and your laundry waste water. This will be filtered and you can use it to water the garden, wash the car, hose down the paths and so on.

Energy efficiency

All new homes built in Australia must achieve a six-star rating for energy efficiency (see Chapter 11). The star rating for a home is between 0 (poor performance) and 10 (requiring virtually no energy to be used for heating or cooling).[5] Meeting the six-star standard – or better – will help make your home more sustainable, but you need to plan how to do this at the design stage. Some measures will cost a little more, but you will make up for the cost over a period of time by the savings in water and electricity you make. A very good place to start gathering information is <www.yourhome.gov.au>. Designing your home to use less electricity and water is comparable and may well be simpler than you think.

If your aspect isn't the best or it has limited solar access because of other constraints, like a neighbour's home over shadowing your land or trees that are too close to your building envelope, you can still achieve an energy-efficient home through careful design. Some of these elements include glazing systems on your windows and shading devices that can achieve winter solar gains from windows facing almost any direction while limiting summer heat gain.[6]

The government has many initiatives available, with rebates for solar panels, solar hot water, water-saving devices, rainwater tanks, and low voltage light bulbs. The government website in your state will give you all the relevant information for you to make informed decisions about your home's efficiency.

In the planning stages it is practical to minimise windows on the western elevation of your home as westerly sun heats the glass and makes the rooms extremely hot. Block-out film is easy to install and will reduce heat absorbed through the windows. You can also buy double-glazed windows to use on this side of the house to make the room more comfortable all year round. If your rooms are on the south side and too cold, double-glazed windows also keep the cold out. Installing retractable awnings to help deflect the sun is another practical method.

You may want to increase the thickness of the insulation in the walls and ceiling if you are in a cool area or a really hot area. Weather seals on external doors are a great way to stop cold and hot air coming inside. Smart use of windows to provide cross-flow ventilation, or the use of louvres, lets out the hot air in summer and will enhance your comfort in winter. Don't forget to make sure that your windows and doors are sealed properly. Curtains and blinds will trap air and create a layer of insulation to reduce heat loss on winter nights, and reversible ceiling fans are great to circulate air in rooms.

Energy choices

The planning stage is also when you need to make decisions about what energy sources you will use in your home. Options may vary depending on where you intend to live; for example rural areas may not have access to the natural gas distribution network.[7]

On the other hand, if you are in a rural setting you may find you can easily harness solar energy through solar panels to heat water for showers and use passive solar building techniques to reduce the need to use non-renewable energy sources.

Green power

Being energy smart around the home can offset the extra cost of using Green Power. The prices vary between suppliers, so shop around to find the one for you.

- 100% Green Power costs around $5 extra a week or between 6 and 8 c/kWh.
- 20% Green Power product can cost as little as $1 extra a week.
- 10% Green Power product is supplied free of charge by some energy providers.
- It is important to check if there are any other fees associated with changing to Green Power.[8]

Electricity

The electricity sector is the biggest single contributor to Australia's carbon pollution. In New South Wales, for example, more than 90% of the state's electricity is generated by burning coal, which creates carbon pollution and is damaging to the environment. The average household in New South Wales creates twice as much harmful carbon dioxide as the average family car. Being power smart to reduce electricity use is vital. But you can also help reduce carbon pollution by switching to a cleaner renewable type of energy.

Most electricity retailers have an accredited Green Power option for a slightly higher unit charge. By choosing Green Power, you are supporting the expansion of renewable systems. Contact your electricity supplier or visit <www.greenpower.gov.au>. This website will help you to choose the best way for you to meet your energy efficiency requirements.

Households can generate their own electricity from renewable sources. These can be either grid interactive or self-sufficient, standalone systems. You can also sell your excess energy back to the grid.

Renewable electricity systems are initially expensive to install but have low operating costs and minimum environmental impact. Government rebates are available to offset the initial costs.

Electricity consumption can be reduced through energy efficiency and fuel switching. As energy costs rise and awareness of environmental issues increases, the value of houses with energy-efficient features and renewable energy supply is expected to rise in the market.[9]

Gas

Natural gas is less expensive to use than electricity and produces fewer greenhouse gas emissions. However, gas is also a non-renewable fuel. It is largely used for water heating, room heating and cooking. It can, however, also be used for clothes drying, as a vehicle fuel, and even for refrigeration.[10]

Wood

Wood can be a renewable energy source if it comes from sustainable managed forests. Its use should make no net contribution to greenhouse gases if trees are planted to replace those used, but fossil fuels are usually used in collection and transportation,[11] which kind of negates any benefit of a sustainable resource.

Efficient energy use

Using energy efficiently is the best way to reduce energy bills and environmental impacts while maintaining or even improving comfort levels.[12] Energy efficiency is achieved by the choices you make in planning your home and by the choices you make in how you use energy in your home. There are numerous resources on the inter net to help guide you through this plethora of choices.

Australian climatic conditions place an enormous demand on electricity for heating and cooling. This is reflected in a substantial percentage of the electricity used across Australia. Recent data indicates that space heating and cooling accounts for over 40% of total residential operational energy consumption (Australian Greenhouse Office 2004; Climate Action Network Australia 2005). The next biggest cost is your hot water system. This accounts for 25% of the energy used in an average home. Electricity produced by coal-fired power stations is the main source of energy for households throughout Australia, and this has led to a growing concern about energy conservation and the sustainability of our living standards (Climate Action Network Australia 2005). As a result, achieving better energy efficiency in buildings has become one of the major challenges faced by Australian builders and architects (Gregory 2007).

Insulation

Insulation is usually rated as R1, R2, R2.5 and so on: the higher the number the higher the insulating capacity. You will need to use thicker insulation in the roof and less thick between the walls, although you could use thicker insulation throughout if you feel the need. You can use heavy duty sarking or rock wool in the roof cavity too, and Whirlybirds are excellent for extracting hot air from your roof cavity.

Hot water

Choose the most efficient hot water service and the best energy source to meet your needs. Solar, gas and electric heat pump systems produce far fewer greenhouse gas emissions than conventional electric storage systems, plus you aren't paying for hot water that you haven't used as in the old off peak hot water heater. Gas-boosted solar is the most greenhouse-efficient form of water heating. Government rebates are available, and many energy suppliers will let you pay off your solar-boosted hot water and photovoltaic systems over time.

Locate water heaters close to the areas where hot water is used, as in bathrooms, ensuites, the laundry and the kitchen. Showers usually use the most hot water in a home so it is worthwhile installing WELS three-star rated water efficient showerheads too so you can save on water as well. The WELS scheme ensures they will provide a satisfying shower. Set the thermostat between 60 and 65°C on storage hot water systems, and at 50°C on instantaneous systems.

Turn off the hot water system when on holidays. Put a timer or manual boost switch on the electric booster of solar water heaters and on peak electric storage systems to avoid heating water when it is not needed.[13]

Heating and cooling

Passive design principles can increase comfort and minimise the need for heating and cooling. Using less energy could be as straightforward as incorporating eaves into your house design: these will reduce heat entering through windows in summer when the sun is higher in the sky, and in winter, when the sun is lower in the sky, they allow warmth to enter through the windows.

Use high efficiency gas, electric heat pump or wood heaters for room heating rather than electric convection and radiant heaters. Radiant heaters – the old bar heater – are suitable for bathrooms when used for short periods of time.[14]

Gas heaters and room air conditioners, either ducted or split system, have energy rating labels. The higher the number of stars the more efficient they are. It is advisable to ask the experts when deciding what is best for your home and to make sure you have the right size in HP and watts for the room size that you will heat or cool. We made the mistake of installing a split system that just could not cope with the load, and wasted over $1000.00 in the process.

If you are installing ducted air conditioning it is a good idea to pay a bit more and have three or more zone controls installed. This gives you more control so you are not heating or cooling rooms you are not actually using.

Use ceiling fans instead of air coolers. Ceiling fans can produce a good level of comfort while using very little energy, particularly in muggy areas where air movement is important.

Evaporative systems are an effective way of cooling your home in low humidity areas and can save hundreds of dollars to run compared to ducted air-conditioning. There are reputable retailers who can explain the features and benefits and the savings. They can be installed during the construction of your home. However, these are not effective in humid climates.

If you are doing your part best to use less electricity for cooling in the summer months, it is important to open doors and windows to ventilate your home once the temperature outside becomes lower than the temperature inside. This will help cool your home down and make conditions more comfortable. Our split system (heater/air conditioner) broke down last winter and now we are in the hot summer months we have found that with the new ceiling insulation, courtesy of the Federal government, our ceiling fan and opening the windows around three or four o'clock it is quite comfortable because we are harnessing the cross-flow ventilation to cool the room down.

If you have designed your home to utilise the northerly aspect into your living areas and you shade the western side of your home you have made a good start. Other measures that will see you well on the way to a sustainable home include double-glazing west-facing windows, and providing awnings or planting deciduous trees to provide shade in summer and allow the sun access in winter.

Appliances

Electrical appliances account for about 30% of household energy use.

When purchasing white goods (refrigerators, freezers, clothes washers, clothes dryers and dishwashers) look for the Energy Rating label. This label gives a star rating and annual energy consumption for the appliance. The more stars, the more efficient the appliance.

Did you know that the fridge typically uses more energy in a year than any other appliance? It's responsible for about 13% of the average family's electricity bill. It pays to buy an efficient and appropriately sized fridge.

Choose appliances with a WELS star rating for water efficiency and energy or water saving features, such as clothes washers with cold wash cycles, economy or 'eco' cycles and load size selection.

Avoid using appliances unnecessarily. Dry your clothes on a line rather than in the clothes dryer, and use appropriate load sizes for clothes washers and clothes dryers. Use cold wash cycles and other energy saving features.[15]

Cooking efficiently

There are currently no energy rating labels for cookers to help you choose the most efficient models.

In general, choose gas cooktops rather than electric if you have gas in your area. You can also use bottled gas, just like when you barbecue. Gas cook tops are often cheaper to use than electric, and have more responsive controls and produce less greenhouse gas emissions. A gas cooktop will produce less than half the greenhouse gases of a standard electric unit.

Use a microwave when possible rather than an oven, as they use less than half the energy. Fan-forced ovens are about 30% more efficient than conventional units, which can waste up to 90% of the energy used. Look for ovens with high levels of insulation and triple glazed, low-e coated windows. Some electric ovens can be divided into compartments for cooking small items.[16]

Lighting

Fluorescent or compact fluorescent lamps and LEDs are energy efficient, using about one quarter of the energy of incandescent bulbs, and long lasting. Avoid using low-voltage down-lights for general lighting as they are not energy efficient. Compact fluorescent replacements for down-lights are becoming available.[17]

Turn off lights when not needed. Use floor or table lamps; they are great for setting the mood and usually less hot in summer. Use timers or sensors on outdoor security lights, and consider using solar lighting for outdoor areas. We love our solar lights we installed last year – they bring a certain ambiance to the garden.

Reducing stand-by energy consumption

Standby energy is drawn by some electrical equipment when it is not actually being used, such as when the TV is turned off with the remote control rather than with the switch on the set or at the wall.

Standby energy use can account for 10% or more of household electricity use, so it pays to be aware of the standby energy use of electrical equipment such as TVs, videos, clocks, computers, faxes, microwaves, security systems, battery chargers and power packs.[18] There are devices that can be installed so you can switch off all your electronics at the same time; you will need one of these per area. You may be pleasantly surprised at how much you can save by making some simple changes such as those described.

Useful websites

Contact your state or territory government or local council for further information on passive design considerations for your climate: <www.gov.au>.

This chapter was based on the websites below. They are a good place to start if you need more information on how to save energy and do your bit for the environment. Once you start looking into it you will see you don't have to make too many sacrifices as it is easier to fit these devices while your home is under construction rather than it is to retrofit them later.

Australian Bureau of Meteorology: <www.bom.gov.au/climate/environ/design/design>
Department of Climate Change and Energy Efficiency: <http://www.climatechange.gov.au>
Centre for Design, RMIT University: <www.cfd.rmit.edu.au>
Department of Environment and Climate Change NSW: <www.environment.nsw.gov.au>
Department for Planning and Infrastructure, Western Australia: <www.dpi.wa.gov.au>
Environmental Protection Agency, Queensland: <www.epa.qld.gov.au>
Institute for Sustainable Futures, University of Technology Sydney: <www.isf.uts.edu.au>
Sustainability Victoria: <www.sustainability.vic.gov.au>
Western Australian Planning Commission: <www.wapc.wa.gov.au>
www.buildingcommission.com.au
www.climatesmart.qld.gov.au
www.energystar.gov.au
www.livingthing.net.au
www.yourhome.gov.au

Further reading

Hollo N (2011) *Warm House Cool House: Inspirational Designs for Low-Energy Housing.* 2nd edn. Choice Books, Marrickville
Wrigley D (2004), *Making Your Home Sustainable: A Guide to Retrofitting.* Scribe, Carlton North, Victoria.

7

Building materials

The choice of building materials is just as challenging as the choice of floor plan. The materials your home is made from will influence its longevity and the security, comfort and health of your family, and has an impact on our environment. There are so many mediums in which to build a home these days, and once you start researching options the decision may become even harder.

Recycled materials

You may consider using recycled products in building your home. There is a growing move towards recycling building materials as this can minimise our carbon footprint. Construction and demolition materials can constitute up to 40% of the waste generated in Australia each year. It makes good sense to recover and recycle as many of these materials as possible.

Materials that can be recycled include masonry, concrete, construction aggregate, bricks, tiles, timber, glass, insulation material, metals, trees, organic matter, plastic fittings, bath, toilets, sinks, pavements, roads, car parks, driveways, pipe bedding, floorboards, reconstituted panel boards, doors, furniture, fencing, mulch, crushed glass reused as compaction fill.

Although it is doubtful you will be allowed to use recycled materials in a new housing estate, during our research we came upon someone who recycled old car tyres in his concrete slab rather than buy new waffle poly pods, thereby reducing his carbon footprint and saving money. How easy was that!

During the construction process, an excessive amount of waste is often generated. This can be in the form of materials that need to be cleared from the site, such as trees and soil, leftover materials and off-cuts, and post-consumption materials. Removal of these materials incurs labour, storage and transportation costs, as well as landfill dumping fees. The cost of removal and disposal can be significant. An on site study by Fletcher Construction found that waste volumes could be greatly reduced through educating the workforce about recycling and implementing a simple waste management strategy.

Minimising the waste produced by the construction and deconstruction of the built environment has a number of benefits. Not only does construction and demolition waste take up valuable landfill space but it is often comprised of valuable resources.

Substantial monetary savings can be attained by reducing and recycling waste generated by the building and construction industry. To find out where to purchase recycled materials just do a web search – it is fascinating. Some places have 100-year-old timber sleepers! All you have to do is use your imagination on how you can use them.

Environmentally friendly building materials

From the Victorian MBA website you can access a draft document called Eco selector <www.auroraliving.com.au>. Click on the resources tab and the first PDF is Eco selector, which may help you to choose environmentally preferable building materials.

Eco specifier is a guide to eco-preferable products and materials for the construction industry. It is produced by a not-for-profit collaboration that aims to help create a more sustainable physical environment by increasing the use of environmentally preferable and healthy products, materials and design processes.

Eco Recycle Victoria have produced the Resource Efficient Builder guideline to provide advice on how builders can be more efficient, produce less waste, recycle more, and have less impact on the environment <www.sustainability.vic.gov.au/resources/documents/Waste_Guidelines1.pdf>.

To improve the sustainability of the building environment you might source materials such as non-toxic, low embodied energy building materials and using materials in such a way that they can be re-used or recycled. Using materials from low environmental impact sources, for example recycled floorboards, or value-added plantations rather than native forests, is an important sustainable practice.

There is also an additive that you add to paint to help save energy. Thermilate is an additive that makes paint insulate and claims to cut energy consumption by up to 40%. Ceramic micro-spheres create a thermal barrier that refracts, reflects and dissipates heat and can be used on the interior to keep heat in during winter.

The foundations

It is obviously very important to have a really good base for the construction of your home. Without this the house may become unsound and could result in damage to the structure of your walls, ceilings and roof.

As explained in Chapter 5, the type of footing (e.g. slab type or peering system) you use for your home may be determined by the slope of your block of land, the load the foundation has to bear including the materials you choose, the type of soil, salinity, and if you have mine subsidence. Usually an engineer's report will tell you what type you must have. Australian Standards 2870 slab will apply to all new constructions from 2012: see Chapter 5.

Concrete slab construction centres on footings poured for support. Reinforcing materials set within the concrete help prevent cracks and settling.

There are several different types of footings you may use. You will need to consult with an expert to decide which one is the best for your circumstances and consider all your options. From our experience most volume builders use waffle pod slabs as it is the least expensive option; however, you may not be able to use a waffle pod slab if you have a steep site or if you have a sheer drop off the side of a hill or rock face. In these cases you may have to have a pole house, which is built up off the ground using poles, or you may use suspended concrete: these are outlined below. In flood-affected areas or areas with drainage problems the house must be raised up off the ground so that water can flow freely under the house and not cause internal damage.

We should point out here that every builder has a different name for each of the footing types, and this makes for confusion. Regardless of the name they all come under the umbrella of footing systems. Since there is no other way to describe the types of footings, we have referred to <www.yourhome.gov.au> extensively to be clear and concise.

Stiffened raft slab

The stiffened raft slab it is a slab on the ground with edge beams or piers embedded into the soil. Internal beams and steel mesh are laid before the concrete is poured; this is performed in one operation.[1]

Waffle raft slab

In a waffle raft slab or waffle pod the slab sits on the ground and is not entrenched into the ground. The slab ribs are formed on top of the ground using a grid of polystyrene void forms that are laid out on a levelled area. There is also the ribbed raft pod system, which is designed for single-storey brick veneer dwellings.

The waffle raft slab system uses less concrete than the conventional method of edge beam and internal beam trenching, so the need for labour intensive digging is eliminated. Since the void formers are waterproof, this system can be used even during wet weather conditions.

Another reason the waffle pod slab is useful is because it is made up with polystyrene and void formers. These voids improve the energy efficiency function, significantly reducing heat loss through the floor. The thermal mass stores and re-releases thermal energy, which in turn can help regulate indoor comfort by radiating or absorbing heat. Thermal mass is useful in most climates, and works particularly well in cool climates and climates with a high day–night temperature range.

To be effective, thermal mass must be used in conjunction with good passive design. Design slabs to absorb heat from the sun or other sources during winter. Heat can be stored in the slab and re-radiated for many hours afterwards. In summer, it allows the slab to be exposed to cooling night breezes so that heat collected during the day can dissipate.[2]

Strip footing slab

The strip footing slab consists of a slab supported by strip footings. Strip footings, also called edge beams or drop edge beams, are a reinforced concrete beam installed into soil and laid around the entire outside of the house. They require a minimum specified depth of concrete to ensure that they have adequate strength. A footing slab requires two separate concrete pours.[3]

One major advantage of footing slabs is that they can be used well on sloping blocks. They require simple formwork, require minimal excavation, and do not require internal beams. This type of slab can be used in conjunction with a waffle pod slab for sloping sites.

Column or stump sub-floor

The column or stump sub-floor is useful on sloping sites. This footing is used for the vertical support and the transfer of building loads to the foundation. Stumps are used to support timber-framed houses, either weatherboard or brick veneer, for which they are currently the most cost-effective option. Materials commonly used for stumps are timber, concrete, and steel.

A sub-floor system consists of stumps, bearers, joists and sheet flooring. The beams that connect directly to the stumps are called bearers, and the materials used that sit on top of the bearers are called joists.

Once the floor frame has been constructed, plumbing and other services can be installed. Tongue-and-groove sheet flooring can be glued, nailed or screwed to the joists. And then wall framing will commence.[4]

Suspended concrete slab

As the name implies, the suspended concrete slab is raised off the ground and has an accessible sub-floor area. For a typical residential situation the slab is around

200 mm thick and is supported by external sub-floor walls of brick or concrete block and by free-standing brick or concrete piers. The suspended slab can also be supported on a Bondek® steel composite system or a suspended slab which has two layers of steel mesh reinforcement.[5]

Pier and beam foundation

Pier and beam foundations are commonly constructed of reinforced masonry (brick or concrete block) supported by individual, reinforced-concrete pad footings or by continuous, reinforced-concrete spread footings.[6]

Pole construction (post and concrete)

A pole house is useful in areas where you want to take advantage of views with minimal site excavation and alteration to terrain. This is always good for the environment, and may be necessary if you have a steeply sloping site. In a pole platform, poles are used to support a level frame on which a conventional house frame is built. A pole frame uses poles extended throughout the house right through to roof level to support the entire building.

For this type of building footing a hole is dug into the ground fairly deeply then a pole is placed into the hole and concrete is poured in.[7] There is a bit more to it though: you will have to consider things such as how long the poles have to be, the right height for the poles, how far down the pole have to be, if it is better to use steel than wood. The best thing to do is to talk to an expert.

Concrete slab finishes

For the thermal mass of a concrete slab to work effectively, it must be able to interact with the house interior. Covering the slab with finishes that insulate, such as carpet, will reduce the effectiveness of the thermal mass.[8] The technical manual available from <www.yourhome.gov.au> provides information on many different materials.

Tiles

Tiles fixed by cement or cement-based adhesives are commonly available in many colours, sizes and patterns. (If thermal mass is to be utilised, avoid rubber-based adhesives due to their insulating effect.) Darker colours with a matt surface work better than light shiny finishes. Choices include ceramic tiles, slate tiles, porcelain tiles, terracotta tiles, pavers and bricks.[9]

Ceramic and porcelain tiles are conductors of heat and tend to absorb the heat from your feet, so they feel cold.

Polished concrete

Polished concrete and coloured concrete can be used in either steel trowel or burnished finishes, to achieve various results. It may be advisable to use experienced specialist contractors to carry out this work. Colours can be applied as oxides in the mix, or as 'dry shake' pigments applied to freshly screeded concrete and then trowelled in, or by chemically staining the concrete.[10] This finish can be highly glossy or matt, and looks spectacular.

Floating timber flooring and laminate floors

Laminate flooring is a tongue-and-groove interlocking flooring system that floats on top of the existing sub-floor. A relatively new product in the Australian floor-covering industry, it has a core of hard material with a laminated printed layer and a special backing that are secured to the core and then saturated in plastic-type resin called melamine.

Laminate floor is not attached to the floor underneath. A special polyurethane padding is laid down prior to the new laminate flooring being installed, which prevents the glue from sticking to the sub-floor. The sub-floor could be a new wood sub-floor, a concrete slab, an existing vinyl floor, or any other existing flooring type.

It is very easy to install the floating floor. All that is required is a bead of specially formulated water-resistant glue to be placed between the tongue and grooves of every plank to hold the planks together and to seal all the edges of the planks from moisture.[11]

Timber flooring and hardwood floors

Hardwood is a classic flooring material adaptable to every decor from wildly contemporary to formal. The stylish look of hardwood floors can add warmth and charm to any room in the home. The natural characteristics of wood add depth and a rich appearance that many other types of floors try to duplicate. As the demand for hardwood floors has grown so has the manufacturers' ability to produce better quality finishes and superior construction techniques. With these advancements wood floors can now be installed throughout the home and over a wide variety of subfloors.

While not technically a wood, bamboo is also a popular material for modern floors.[12]

Cork

Cork is a natural insulator of heat and so feels very warm underfoot. Cork is naturally an excellent insulator of sound, making it very quiet to walk on. As cork is simply millions of air cells trapped in a honey-coloured casing, walking on cork is like walking on a natural air cushion. The natural characteristics of cork make it an excellent choice for residential flooring.[13]

Vinyl

Vinyl flooring can bring stylish designs and practicality to any room. Vinyl surfaces are easy to maintain, comfortable underfoot, and suitable for many domestic and commercial applications. Vinyl is available in many attractive patterns and effects, from traditional colours and patterns to slate, marble, timber and cork effects. Vinyl tiles can also offer an alternative to traditional ceramic tiles. You can even use vinyl to create your own individual designs.[14]

Parquetry

Parquetry is a geometric mosaic of wood pieces used for decorative effect. The two main uses of parquetry are as veneer patterns on furniture and as block patterns for flooring. Parquet patterns are entirely geometrical and angular – squares, triangles, lozenges. The most popular parquet flooring pattern is herringbone. (The use of curved and natural shapes constitutes marquetry rather than parquetry.)

Timbers contrasting in color and grain, such as oak, walnut, cherry, lime, pine and maple, are sometimes employed, and in the more expensive kinds the richly coloured mahogany and sometimes other tropical hardwoods are also used.[15]

Frames and exterior walls

The façade of a building is often the most important from a design standpoint, as it sets the tone for the rest of the building and enhances the appearance of the streetscape. Particularly if you are building in an estate there may be restrictions on your choice of external materials.

The face of your home and the way it is presented is vitally important to the overall finish of your home, and good choices will add substantial value.

The choices of external wall material for houses are usually face brickwork, rendered and painted brickwork/concrete blocks, bagged and painted brickwork/concrete blocks, stone, or a combination of these materials. Second-storey walls may feature light-weight materials such as fibre-cement sheeting, fibre-cement or timber weatherboards. These can be finished with a texture-coated paint finish or simply painted with a complementary colour.[16] The technical manual available from <www.yourhome.gov.au> provides information on many different materials.

Mud brick

You probably would not make your own mud bricks – it would take several months to make enough bricks to build your home, and where would you store them? However, it is possible to make your very own earth building. Earth can be used to construct walls, floors, roofs and even furniture, fireplaces and ovens. It is a building technology with an 11 000-year-old history and tradition, and is utilised worldwide. Today it is estimated that one-third to one-half of the world's population is housed in earth homes.

The common feature in all earth building techniques is that the earth material is subsoil that is composed of clay, silt, and sand. The clay is the binder or cementing ingredient. Mud bricks are joined with a mud mortar.

Basic mud bricks are made by mixing earth material with water, placing the mixture into moulds. After the initial drying the moulds are removed and the bricks are dried slowly in the open air, using only the heat of the sun. Straw or other fibres that are strong in tension are often added to the bricks to help reduce cracking.

Virtually all the energy input for mud brick construction is human labour and the sun. After a lifetime of use, the bricks break back down into the earth they came from. The most effective use of mud bricks in building healthy, environmentally responsible housing comes from understanding their merits and accepting their limitations.[17]

Mud brick embodies the material they are made from. They are usually reddish, yellowish or grey, depending on colour of clays and sands in the mix. This type of building material lends itself to many modalities, and can be used to build walls, vaults and domes. You are not restricted to a rustic look as it can be smoothed in the same way as cement render if that's the look you prefer.

Mud brick and adobe walls can provide moderate to high thermal mass, meaning they are slow to heat up and slow to cool down. Earth building has a small carbon footprint because it uses the sun for drying and the bricks are usually hand-made.

Bear in mind that the average home uses 20 000 bricks – you will have to have a very big working bee to make your own bricks! There are, however, retail outlets where you can buy them.

Straw bale

Straw has been used as a building material for centuries because of its structural integrity, both for thatch roofing and mixed with earth in cob and wattle and daub walls. Straw bales were first used for building over a century ago, shortly after the invention of baling machines.

Straw is an annually renewable resource that is mostly wasted: each year, millions of tonnes of straw are burnt or left to rot. If all the rice straw in the Riverina of New South Wales was baled and used to build an average 200 m^2 straw bale house, then we could build 48 000 straw bale homes on an annual basis. There are grain crops grown in most parts of Australia and New Zealand, so this building material is available almost everywhere with minimal transportation costs. Straw from wheat, oats, rye, rice and barley can be used to make bales suitable for building construction.

Finished straw bale walls are usually rendered with cement or earth so that the straw is not visible. The final appearance of rendered straw bale can be very smooth and almost impossible to tell apart from rendered block work, or it can be more expressive and textural.

The structural capacity of straw bale construction is surprisingly good, and you can easily build a two-storey home. Most straw bale construction uses a timber or steel frame to comply with current building codes.

Wool brick!

Wool bricks are an innovative product that could help to reduce our carbon footprint.

Ask yourself: do you have more wool, clay and seaweed than you know what to do with? Here's a solution: make really strong bricks. Researchers in Spain and Scotland say they've done just that. In experiments conducted at the University of Seville in Spain and the University of Strathclyde in Glasgow, Scotland, researchers added wool fibres to the claylike soil used to make bricks, and then threw in alginate conglomerate, a polymer made from seaweed, according to a study published in the journal *Construction and Building Materials* (Galán-Marín *et al.* 2010a).

The bricks with wool were 37% stronger than conventional bricks and were more resistant to cracks and fissure, the researchers reported. Wool bricks are also energy savers as they're made without firing.

'This is a more sustainable and healthy alternative to conventional building materials such as baked earth bricks and concrete blocks,' the study's authors, Carmen Galán and Carlos Rivera, said.

The bricks aren't going to force anyone to give up their kilts or sweaters. Scotland's sheep farmers produce more wool than its textile industry can use, the researchers say.

The perception that straw bale is cheaper than conventional building technology is incorrect. Straw bale building is competitive with other forms of building but it can be more expensive, depending on the design and complexity of the building itself. However, in terms of sustainability it is far better for the environment than traditional brick veneer.

Straw is tough and fibrous and lasts a long time if kept dry. The bales themselves have very low thermal mass, being composed, by volume, mostly of air. However, the cement and earth renders typically used on straw bales result in finished walls that have some appreciable thermal mass in the thin masonry 'skins' either side of the insulated straw core. With the use of earthen renders a thick render skin of up to 75 mm can be achieved, providing significant thermal mass. In addition, finished straw bales need to be waterproofed externally and, as with all homes, all wet areas must be comply with the building code. If you are considering straw bale construction in a bush fire area it must also comply with relevant building codes. Straw bales demonstrate excellent insulation properties, in fact possibly the most cost-effective thermal insulation available, not to mention sound insulation, and with the typical coating of cement render they have a low fire rating.[18]

Light-weight timber

Wooden structures have long been used in all kinds of building. Lightweight timber construction has a long history and is the most common house construction type in Australia.

One of the key advantages of timber is that it provides an adaptive material for use in all climatic zones. The lightweight timber house can provide cost-effective and flexible design options.

Timber frames can support internal and external walls, floors and roofs. A variety of non-structural claddings, linings and finishes can be used, such as weatherboards and timber fibre products, or non-timber products such as brick veneer, fibre-cement sheet, or metal.

Timber framing is a versatile, environmentally sustainable method of construction. Advantages of lightweight timber framing include that timber is an easy material to handle and work with, and the framing can be adjusted easily for future alterations or additions to your home. Timber framing is a low-energy system of construction. Framing can be erected quickly; an owner builder acting as labourer and a carpenter can build the frame of a three-bedroom home in about one week.

Timber houses can range in appearance from the ultra-modern to the traditional weatherboard house. Depending on the cladding used, the appearance may express the timber construction or disguise it – most timber framed houses in Australia are finished in brick veneer.

Timber is an organic material and deteriorates due to weathering. The main way of preventing weathering is protection of the timber surface. This may be achieved by appropriate design detailing, so that the timber remains dry or sheds water quickly. It may be achieved by treatment with an appropriate surface coating of oil, varnish or paint. Such coatings on external timber components of buildings generally need replacing every 5–7 years or sooner in more extreme conditions.[19]

All homes built with timber frames must have a termite barrier, or spray annually to prevent termite infestation.

Cladding

Cladding is an external skin applied to a steel or timber frame. There are some very exciting cladding products you can use, either in conjunction with brick veneer or stand-alone. Each has its unique installation system. It is highly recommended that when you find one that you like you read the product technical guide, as some are not so easy to install and are labour intensive. Light-weight claddings are generally easy to handle and easy to cut during construction.

Cladding treatments have their own distinctive attributes with minimal maintenance once they have been installed properly. A consideration for those building in fire-prone areas is that many claddings have attributes like fire resistance, meaning they can withstand a certain amount of exposure to fire – but not that they won't ignite after a certain period of exposure. The fire rating standard will be in the specifications sheet. Claddings withstand or are impenetrable to water, and most are termite resistant.

There is a huge variety of cladding systems that will add value and appeal to your home. We have included just a few of the cladding systems available but there are many more to choose from.

James Hardie has a product cladding series called Scyon. This is an advanced lightweight cement composite with heavy-duty performance. Scyon comes in many profiles and sizes.

Metal Cladding Systems manufacture a profiled design in many different coloured metals. Metallic sheet metal finishes and standard sheet metals such as copper, aluminium, titanium, zinc and stainless steel are available.

CSR Cemintel Scarborough Weatherboards are suited to many residential external cladding applications including upper and lower storey additions, composite construction, over-cladding of existing walls, gable ends, infill panels and feature panels.[20]

Stone cladding or stacker stone is great feature to add to your exterior façade or as an interior feature wall. There are a number of retail outlets and suppliers in all states. It usually costs around $65.00 per square metre. If you find one that looks inexpensive it may not be real stone and instead be made up of composite materials, so be careful and do your homework before you buy.

Steel frames

Steel is the best performing building product in strength:weight ratio. It is compatible with most other building products, does not shrink, crack, rot or burn, and is 100% termite resistant.

The wall and trusses are pre-fabricated, ensuring a very fast building method. The light weight means that no heavy lifting equipment is required: two men can easily carry a wall 6 m long. Less weight also means less movement of the mass of the building. Walls are thinner, therefore saving building mass overall and so foundations can be lighter, and structural beams can be smaller and lighter.

Steel is 100% recyclable. Currently in Australia, it is estimated that of all steel products from building demolition 82% are recovered, ranging from 95% recovery for structural steel (world-class recovery) to 70–80% for reinforcing steel.[21]

Clay bricks

Clay bricks are made from selected clays that are moulded or cut into shape and fired in ovens at temperatures of around 1200°C. The firing process transforms the clay into a building component with high compressive strength and excellent weathering qualities. These attributes have been exploited for millennia to build structures ranging from single-storey huts to enormous viaducts.

The bricks are readily available, mass produced, thoroughly tested modular building components. Their most desirable acoustic and thermal properties derive from their relatively high mass. Clay bricks are generally affordable, require little or no maintenance, and possess high durability and load-bearing capacity. The use of clay brickwork is informed by extensive Australian research, manufacturing and construction experience.

Clay brickwork is available in a wide variety of natural colours and textures derived from fired clay used in combination with cement mortar joints of various colours and finishes. Bricks remain stable and colour-fast, and do not need to be rendered or painted. Clay brickwork is most commonly used uncoated to display the richness and texture of the material.

A high thermal mass means that brick responds slowly to temperature changes. The thermal resistance of clay brick veneer or cavity walls can be greatly enhanced by adding foil or bulk insulation.[22]

Double brick

Building in double brick adds a considerable amount to the cost of building, not to mention the labour costs. Added to this, the walls heat up slowly and stay warmer for long periods, so a double-brick building can be hotter in summer. Double brick walls need to be well insulated to prevent heat transfer to the interior of the home during summer and help to retain heat during winter.

Brick veneer

Brick veneer is the typical brick construction. The walls consist of a single external layer of brickwork, with a lined stud frame inside. These walls have less thermal mass than double brick walls and therefore respond more quickly to temperature changes. Homes with brick veneer walls are better at cooling down during extended periods of hot weather, making conditions more comfortable at night during summer. Brick veneer walls are also easier to insulate.

Reverse brick veneer walls have the brickwork inside and lightweight frame and cladding outside. This has the advantage of providing the thermal mass on the inside of your home, keeping you cooler in summer and warmer in winter.[23]

Autoclaved aerated concrete

Autoclaved aerated concrete (AAC), sometimes known as autoclaved cellular concrete (ACC) and marketed as Hebel PowerBlocks, was invented in the mid-1920s by the Swedish architect and inventor Axel Eriksson. The use of AAC in Australia is not yet widespread, but autoclaved aerated concrete blocks have been used in Europe for more than 50 years. It is a light-weight, precast building material that provides structure, insulation, fire resistance and mould resistance. AAC products include blocks, wall panels, floor and roof panels, and lintels.

AAC has since been refined into a high thermally insulating concrete-based material used for construction both internally and externally. One of its advantages is its quick and easy installation, since the material can be routed, sanded and cut to size on site using standard carbon-steel band-saws, hand saws and drills. Its easy workability allows accurate cutting that minimises the generation of solid waste during construction. Its excellent thermal efficiency means that AAC can eliminate the need to be used in combination with insulation products that otherwise increase the environmental impact and cost of construction.

Even though regular cement mortar can be used, 98% of the buildings erected with AAC materials use thin-bed mortar that is applied with a toothed trowel, although more conventional thick-bed mortar can be used.

The closed air pockets contribute to the material's insulating properties and also its aerated nature. Although there is no direct path for water to pass through the material, the porous nature of the material can allow moisture to penetrate the material to some depth. Damp proof course layers and appropriate coating systems prevents this happening. AAC will not easily degrade structurally when exposed to moisture, but its thermal performance may suffer.

AAC material can be coated with a stucco compound or plaster. Materials such as brick or vinyl siding can also be used to cover the outside of AAC materials. A number of proprietary acrylic polymer finishes are available that, when applied over sand and cement render, provide a very durable and water resistant coat. This is similar to the acrylic polymer coatings applied before tiling in wet areas such as showers. The manufacturer can advise on the appropriate coating system, surface preparation and installation instructions to give good water-repellent properties.

The introduction of Australian Standard AS3959-2009 'Construction of buildings in bushfire-prone areas' (see Chapter 12) presents new challenges to building designers, builders and homeowners when making the important decision on building materials, particularly for external walls. Hebel blocks and panels are non-combustible and have been tested at CSIRO to achieve excellent fire ratings. This unique material delivers the right balance between insulation and thermal mass and even under extreme heat, it will not explode. All of Hebel's solid and lightweight framed systems meet or exceed the requirements in all of the new six Bushfire Attack Levels (BAL) including BAL-FZ (Flamezone) for external walls, floors, subfloor supports, decks and verandahs. (See Table 12.1 for an explanation of BAL.)

Hebel PowerBlock homes provide frameless load-bearing walls and when constructed in accordance with the masonry code and Hebel's PowerBlock design and installation guides, comply with all six BAL requirements with a Fire Resistance Level (FRL) of 240/240/240 – far higher than the standards require. Hebel PowerWall – which uses 75 mm Hebel steel-reinforced panels on a framed construction – achieves an FRL of 240/180/180 – still exceeding the requirements even of BAL-FZ.[24]

Windows

Well designed windows and skylights provide natural light while keeping winter warmth in and summer heat out.

The Window Energy Rating Scheme (WERS) ranks windows for their energy performance in typical housing anywhere in Australia. It will tell you whether a given window is suitable for the climate or not.

Rating of a window for energy performance starts with establishment of basic solar, thermal and optical properties of the glazing unit and window frame. These properties are determined by a combination of laboratory measurements and computer simulations.

1. Identify the climate classification for the site.
2. Follow the window selection guidelines for climate type and identify generic window types that might be suitable.
3. Compare the WERS star ratings for the suitable generic windows with products recommended by local distributors and make a selection based on cost and performance.

WERS ranks windows in terms of their whole-house energy improvement when compared to the base-case window (a single glazed clear window with a thermally unbroken aluminium frame). The rankings are then used to generate star ratings for cooling (summer and solar control performance) and heating (winter performance).[25]

The technical manual available from <www.yourhome.gov.au> provides information on different options for windows.

The roof

Roof materials can be pre-finished concrete tile, terracotta tile, or non-reflecting coated metal. The technical manual available from <www.yourhome.gov.au> provides information on the different options.

Concrete roof tiles come in many colours and profiles. Some have the colour just on the outside and the better ones have the pigment colour going all the way through to the inside. These do not fade as quickly through the life of the tile.

Terracotta tile used to be very popular, but with the variety of profiles and colours available most people prefer to match their roof colour to the downpipes and guttering and to a certain degree the window frames.

Steel roofs are also becoming more popular. With slim profile corrugation and the many choices of colour it is a really attractive alternative to the use of tiles.

References and useful websites

Hyde R (2000) *Climate Responsive Design: a study of buildings in moderate and hot humid climates*, E and FN Spon, United Kingdom

Lawson B (1996) *Building Materials, Energy and the Environment: Towards ecological sustainable development.* RAIA, Canberra.

Mithraratne N, Vale B and Vale R (2007) *Sustainable Living: The role of whole life costs and values.* Butterworth-Heinemann, Cambridge.

www.environmentdesignguide.com.au (for information about the environmental impact of building materials)

www.gov.au (to contact your state or territory government or local council for further information on building sustainability and energy efficiency)

www.yourhome.gov.au (to download a technical manual that provides information on many different materials)

Recycled materials

Clean Washington Center (undated) 'Construction & Land Clearing Market Assessment.' Fact Sheet.

Environment Australia (1998) *Waste Wise Construction Program Handbook: Techniques for reducing construction waste.* Department of the Environment, Canberra.

Environment Australia (undated) 'Construction and Demolition Waste.' Information sheet. Environment Protection Group.

Government of South Australia (undated) 'Recycled Pavement Materials.' Transport SA brochure.

McKeller J (1991) *The Recycling of Building and Demolition Materials: a major opportunity to reduce waste going to landfill.* Recycle 91 Conference Papers, 9–10 September 1991. <www.kesab.asn.au/factsheet9.html>

Stickles G (1994) Construction company finds being green pays dividends. *Business Review Weekly*, 4 July.

Foundations

www.builders-directory.com.au/understanding-house-structure.shtml

www.choiceconcreting.com.au (to see pictures of the different slab types under construction)

www.concrete.net.au

www.homeimprovementspages.com.au (this site lists many tradespeople and resources throughout Australia.)

www.treeworld.info/attachments/f6/14648d1269211700-draft-as2870-residential-slabs-footings-comment-dr-2870-residential-slabs-footings.pdf (this pdf is full of technical data for those who like that sort of stuff)

Mud brick

BDEP *Environment Design Guide*, RAIA: <www.environmentdesignguide.net.au>

Biano A (2002) *The Mud Brick Adventure*, Earth Garden Books, Trentham, Victoria.

CSIRO (1995) *CSIRO Australia Bulletin 5: Earth Wall Construction*, CSIRO, North Ryde, NSW.

Earth Building Association of Australia: <www.ebaa.asn.au>
Simmons G and Gray T (Eds) (1996) *The Earth Builders Handbook*, Earth Garden Books, Trentham Victoria.
www.yourhome.gov.au

Wool brick

Galán-Marín C, Rivera-Gómez C and Petric J (2010a) Clay-based composite stabilized with natural polymer and fibre. *Construction and Building Materials* **24**(**8**), 1462.
Galán-Marín C, Rivera-Gómez C and Petric-Gray J (2010b) Effect of animal fibres reinforcement on stabilized earth mechanical properties. *Journal of Biobased Materials and Bioenergy* **4**(**2**), 121–128.

Straw bale

The Australian Straw Bale Building Association: <www.ausbale.org>

Timber

Australian Hardwood and Cyprus Manual (2003), <www.hardwood.timber.net.au/about.htm>
Environmental Benefits of Building with Timber (2004), <www.tastimber.tas.gov.au/species/pdfs/Environment%20Benefits%20of%20Timber.pdf>.
Forsythe P (2005) *A Review of Termite Risk Management in Housing Construction.* Forest and Wood Products and Research Development Corporation. <www.timber.org.au/resources/A%20Review%20of%20Termite%20Risk%20Management%20in%20Housing%20Construction.pdf>
Gray A and Hall A (Eds) (1999) *Forest Friendly Building Timbers.* Earth Garden Books, Melbourne.
Low D (Ed.) (1995) *The Good Wood Guide.* Friends of the Earth, Melbourne.
National Timber Association (2001) *Environmentally Friendly Housing Using Timber: Principles.* National Timber Association <www.tastimber.tas.gov.au/species/pdfs/Enviro%20Friendly%20Housing.pdf>.
www.timber.org.au

Cladding

www.cemintel.com.au
www.metalcsystems.com.au
www.scyon.com.au

Steel frames

www.aip.gov.au

Clay bricks

Energy Smart Housing Manual, Victorian Government: <www.sustainability.vic.gov.au/resources/documents/ ESHousingManualCh061.pdf>

Think Brick Australia: <www.thinkbrick.com.au>

Autoclaved aerated concrete

Aroni S, de Groot GJ, Robinson MJ, Svanholm G, Wittman FH (Eds) (1993) *Autoclaved Aerated Concrete – Properties, Testing and Design.* RILEM Technical Committee, FN Spon, London.

Comité Euro-International du Beton (1978) *Autoclaved Aerated Concrete: CEB Manual of Design and Technology.* The Construction Press, Lancaster, UK.

Hebel: <www.hebelaustralia.com.au>; also <www.hebel.co.nz/about/hebel%20 history.php>

Staines A (1993), *Australian House Building the Easy Hebel Way.* 2nd edn. Pinedale Press, Caloundra.

www.buildinggreen.com/auth/productsByCsiSection.cfm?SubBuilderCategoryID=685

www.csr.com.au

www.foamed.com.cn

8

Tradespeople

The owner builder has guidelines to follow for preventing conflicts and how to run a smooth business-like transaction with tradespeople and services.

There are tradespeople and businesses in Australia that want the owner builder's business. They find the owner builder's business attractive because of the higher possibility of being paid quickly when working on other commitments. This confirms there is room for the owner builder to negotiate on price, but be realistic and be prepared to pay standard rates. A tradesperson who feels the price is too low will not feel obliged to do their best. Respect any extra effort they show, and small rewards for extra effort may not go astray.

We stress the importance of talking to the tradespeople in an amicable way about your expectations and specifications.

Finding tradespeople

When searching for competent and reliable tradespeople, where do you turn for support? Ask family and friends about their experiences – good and bad – with services they have hired, and check with your state authority if there are any complaints recorded against the company. As a lot of us know, there is the potential for crossing paths with tradespeople who are not operating their business by the rules. However, if you follow some simple steps before committing to work, you can find honest, reliable, and efficient tradespeople.

To avoid insecurity and build trust with a tradesperson check:

- license is registered with state authority
- recent work that proves a high standard of workmanship

A bad experience with a plumber

An acquaintance hired a plumber to work on ground services for a two-storey residential dwelling, but the plumber was not licensed or registered with the state authority.

The plumber's work was inoperable and totally out of specification guidelines. The hydraulics engineer declared the workmanship deplorable. Now there is a dispute between the owner builder and plumber, with the state authority involved. The plumber is refusing to rectify the work and will not part with any payments already issued by the owner.

Not only is there a problem with the plumber, but time and money has been wasted and the project is being held up. If a licence check had been done prior to the hiring of this tradesperson the dilemma could have been avoided.

- current insurance certificates
- fixed address of business
- white card induction approved
- drivers licence and date of birth
- list of references.

We cannot stress enough the importance of checking all tradespeople as above.

Contracts

We suggest having a legally binding agreement when hiring tradespeople. Depending on the amount of the contract price, a signed home building contract could be necessary. You can contact your state authority for details and individual guidelines. Some of the usual details needed when filling in contracts are:

- design plans or specifications
- contract price, warranties
- description of work to be preformed
- date and signatures of both parties
- names and registered licence number.

You are strongly urged to read all clauses printed in the home building contracts, and to be aware of your rights hiring the contractor. A copy should to be given to contractors for their reference.

You can find Plain English home building contracts for sale at your state authority and some news agencies. The contract to choose depends on the cost of work to be carried out, for example Minor Works Contract for work under $5000 and another form for work over $5000.00. The cut-off differs from state to state.

Attempt to have all trades contracts signed and completed before work commences on site, as it is for the protection of everyone involved in the project.

Some suggestions for finding tradespeople are:

- new building areas with tradespeople on site
- Yellow Pages directories
- Local Pages directories
- word-of-mouth referrals
- the list of trades, materials and services in Appendix A.

Once you have selected the tradespeople or contractors, follow the ideas listed in this chapter to ensure reliable and honest completion of work.

Quotations

Making logical decisions as an owner builder is a must. It will be reflected in the final results in quality and cost of the dwelling. Ask the draftsman or architect to email you the custom-made plans. When receiving quotes from several tradespeople for the same job, ask for the same work to be quoted – then it will be easier and fairer to compare them. In other words, if one quote has costed in the pergola on a concrete slab and the other has not, then it will be difficult for you to compare the quotes. Investigate all differences stated in quotes to be safe from hidden costs and dishonest tradespeople trying to win the contract with cheaper quotes.

It is all about communicating with the tradesperson in detail and showing everyone respect. You need to discuss your expectations and provide all plans, designs and specifications to the tradesperson. If there is a special way of doing an

Beware the hidden extras

When we were getting quotes from the concreters for our slab, there were hidden costs. One tradesperson did the right thing and quoted an extra $2000 for piering, as he could see on the structural engineers plans and classification of soil that this would be required. However, we did not want to accept his quote because we had another quote for $2000 less. Once the cheaper tradesperson was given the job, we were hit with an extra $2000 worth of piering on the day he commenced work on our site.

Be aware and listen to the professionals when they are explaining their quotations. Remember: the cheapest quote is not always the best one.

We asked our bricklayer if he could insert our power box into the brick cavity to give a neater appearance so it protrudes only slightly out of the brick cavity. The bricklayer agreed it was a great idea and happily complied, as we had requested this in our initial agreement.

installation or fitting required it should be discussed and documented in the agreement with the tradesperson or contractor. While they would probably welcome the interaction from the owner builder, all requests should be specified in the initial contract.

Deposits and progress payments

Discuss all deposits and progress payments with the tradespeople prior to commencing work. The Home Building Contract will state the maximum percentage a tradesperson can receive as a deposit. Documentation should include an agreed time for completed work and should outline detailed acceptable reasons for delay, for example specified amounts of bad weather, or failure of another contractor or supplier. The progress payments should be written on the contract by both tradesperson and owner builder and specify stages of completion for payment.

9

Insurance

It is important to have a clear understanding of the insurances required by law to be taken out by owner builders, tradespeople, sole traders and partnerships. If a tradesperson is registered as a partnership or sole trader, it is not compulsory for them to purchase and hold a certificate of compensation insurance. Therefore, the individual would claim on the owner builder's insurance.

Before hiring any tradesperson, discuss insurance certificates of currency and subcontractor's statements. This will identify all individuals' insurance cover. Under some insurance policies, not every visitor is covered on site, and in particular people who are not licensed contractors and have been working on the site may be excluded.

The aim in discussing insurance for owner builders and tradespeople is to become aware of the loopholes and to protect the project. An insurance broker can guide the owner builder concerning comprehensive insurance, contract works, public liability, theft, fire, storm policies and so on. Tradespeople can be guided by their state authority for insurance conditions and guidelines.

See also Chapter 14, Contracts.

The owner builder

The following listed insurance cover was for our first house. Not being an expert in this area ourselves, it made sense for us to seek advice from a qualified insurance broker. However, every policy is different and each owner builder will need to purchase and insure on their own merits.

- Public liability $5 000 000.
- Contract works $255 000.
- Damage to existing (neighbours') property.
- Underground services.
- Removal of debris 10% of contract value.
- Professional fees 10% of contract value.
- Escalation costs 10% of contract value.
- Tools $5000.
- Fibre optic damage.
- Owner builder's warranty (only applied if we resold within 6 years).
- Construction plant and equipment.
- Employee personal effects.
- Subcontractors.
- Product liability.
- Bushfire or grassfire.
- Storm.

Usually, the policy will be valid for 12 months or until practical completion of the dwelling. If the owner builder finds the tradespeople in dispute with the insurer, contact the Insurance Council of Australia for assistance to resolve the matter.

The subcontractor

The following lists insurance cover typically required by tradespeople:

- public liability and property damage
- theft of stock, tools and contents
- workers compensation
- home owner warranty insurance
- business vehicle insurance.

Standards and requirements differ from state to state, and insurance policies are to be purchased by tradespeople according to the state authority guidelines. For example, home owner warranty insurance may require a contractor to be licensed with the authority and also require cover for:

- financial loss or damage from non-completion of work for 12 month period
- structural defects
- disappearance of builder
- work for six years from date of completion
- suspension of contractor's licence.

If the tradesperson has a dispute with the insurer, the state authority is available to advise and support the individual on the action to take.

Avoiding problems

To avoid complaints between owner builder and insurer or tradesperson and insurer, do not commence any building works before the insurance policy starts. Otherwise, there will be a risk of not being covered by the insurance company.

Additionally, the smallest error in an insurance policy can create potential problems when making a claim. It is essential to ensure paperwork is correct and accurate. When we received our owner builder insurance policy by fax, we found an error on the front page stated our residential address, not our building address, as the building site.

Unfortunately, there are times where accidents and disasters do happen on building sites. Ultimately, if the relevant insurance has been taken out the outcome will support and protect the insured.

A major trap is underinsuring the project. To estimate the cost of replacement of construction, you need to allow for contractors' labour, construction costs and your own time and labour as owner builder. Owner builders need to cover the project for the amount on the tender or quotation acquired in the beginning of the process and also for anything not included on the tender, such as flooring, blinds, bathroom accessories, and rental accommodation while building.

There is a massive difference between the money you pay for a good insurance policy when building a home and the amount you pay for comprehensive motor vehicle. The cost of the two is similar, but the value of cover is substantially different. For the year we owner-built our home we were covered for over $225 000 on the contracts work insurance and the public liability insurance of $5 000 000, and paid about the same as for car insurance covering only $33 000.

Insurance companies may offer an owner builder the option of a cheaper policy if the owner builder agrees to an excess on the policy. This means that the insurance covers only claims above a certain agreed amount (the excess). Taking out such a policy would be a decision for the individual.

Some insurance companies will not cover existing structures or dwellings during renovation or extensions. It is absolutely necessary to discuss the cover of insurance with any current household insurer. Confirmation in writing from the insurer will be required for the additional works and policies on the existing dwelling.

10

Estimating

How much will it cost to build?

The main question on the owner builder's mind is usually cost. The average amount an owner builder will save on a project would be 20–30% of the price. This can vary, of course, as spending is the responsibility of the purchaser or decision-maker.

A contract builder tendering for a client needs to protect their business with provisional allowances and miscellaneous prices. These allowances can add to $12 000 or more and include:

- stolen goods allowances
- sales consultant commission
- estimator and administration errors
- provisional allowances.

This is why builders don't give you fully exposed prices. Some of these expenses can arise for the owner builder, but you need not necessarily have to face them. Contract builders charge for them regardless, to keep their business from failing through incorrect tenders and commitments, but as an owner builder you can keep the money in your pocket and pay only for necessary expenditure.

An owner builder working out their budget needs to evaluate professional advice carefully. If you go to a contract builder to get a tender on your working drawings, the $12 000 or more of provisional allowances will be included. In addition, the builder would cost his own profit into the tender as well.

However, if you go to a professional unbiased estimator with your plans, the quotation would reflect a more accurate costing as it would not include the allowances above. The best way to request a quotation from an unbiased estimator is to ask for an itemised quotation. However, you would need to include an allowance for errors and oversights anyway, and arrange finance accordingly.

There is an advantage in requesting two or three unbiased estimators to quote on the project when you are creating your owner builder budget plan. Even though the services of a building estimator cost money, it does assist in obtaining a successful financial outcome.

Taking control of the budget

Listed below in five stages of building are the essential materials and services we allowed for when we owner-built our second house. The stages are included to indicate when you will need money available as well as providing a guide to scheduling. The items would vary depending on the type of dwelling, materials, size, land slope, climate, services provided, access, vegetation, streetscape, single or double-storey, soil classification, piering and area, but these should be a useful indication.

The materials we used were raft concrete slab, timber frame, clay-fired brick veneer and Colorbond roof.

Stage 1: Prior to slab

- white card
- owner builder's course
- building permit
- draftsman
- working drawings
- developer's consent
- contour of land
- geotechnical engineer
- surveyor
- home owner's warranty
- Sydney water
- private certifier and council
- NatHERS report
- construction fee
- BASIX if required
- public liability insurance
- site toilet
- excavator and site works

- temporary fencing
- sediment control fencing and signage
- construction insurance
- engineer for structural design and inspections.

Estimated time: 6 months

Stage 2: Preparing slab for pouring, frame and windows

- external drainage
- pest control
- concrete and labour
- frame
- trusses
- frame carpenter
- frame hardware
- sliding doors
- door frames and windows
- plumber and gas rough works
- gas notification
- rubbish removal
- glass blocks and frame.

Estimated time: 4 weeks

Stage 3: Roof and brickwork

- Colorbond roof and labour
- metal fascia gutter
- brickwork supply
- brick hardware
- brickies sand
- structural steel
- bricklayers' scaffolding, bricklayer, rubbish removal.

Estimated time: 4 weeks

Stage 4: Fix out

- wall insulation
- pergolas
- eave material
- eave hardware
- eave carpenter
- electrical rough in

- internal wall and ceiling linings
- metal downpipes
- garage door
- wet area flashing
- plumbing
- private certifier inspections
- carpenter: skirting, architraves, jambs, doors and door handles.

Estimated time: 5–8 weeks

Stage 5: Practical completion and final fix out

- bath, laundry tub
- shower screens and mirrors
- final lock door hardware
- carpenter final fix out
- floor and wall tiler
- kitchen vanity and sinks
- painter
- electrical finish off
- floor sander
- plumber finish offs
- floor coverings
- bathroom toilets
- window coverings
- light fittings, fencing
- retaining walls
- roof insulation
- final pest control

The unforeseen

Although we listed everything we could think of, inevitably there were extra items. Things we had not budgeted for included the following:

Item	Additional cost ($)
Timber hockey sticks for bull nose verandah	1450.00
Excess spoil (leftover dirt)	500.00
Interest rate increases	per annum
Timber poles on verandah (not quoted in frame)	1700.00
Battens for Colorbond roof	1000.00

- house and yard cleaner
- driveway, paths, turf
- kitchen appliances, benchtop
- flyscreens, security doors
- mailbox, clothesline
- hot water system, occupancy certificate
- air conditioning, gardens, water tank.

Estimated time: 8 weeks

Shopping around

Consider shopping around for materials, services and labour. There can be a large variation in prices and this is where the owner builder can save. Compare quotes that have the same items, lengths or brands to give fair comparison between companies. Fair comparisons make the pricing easier to understand and make it easier to decide who to hire for the job.

Wins and losses

In our experience we found some significant differences in quotes and prices as listed below:

Item	Price difference ($)
Metal roof and labour	5000.00
Insurance	1800.00
Engineer	187.00
External plumber drainage	1950.00
Frame carpenter	2009.00
Frame	1585.00
Toilets	321.00
Total saving	**12 852.00**

On the other hand, we went over the budget on the following items and services:

Item	Additional cost ($)
Council fees	550.00
Excess spoil (leftover dirt and debris)	500.00
Laundry tub/sinks	304.00
Total over budget	**1354.00**

Realistic needs

We all have a price in our minds that we want to pay for the project, but we need to try to let go of that and be level-headed. The opportunity to save money when shopping around for materials and labour is real. Budgeting from several tenders or quotes can be important in staying in control.

It is possible something that is necessary for the project may not have been listed in the budget. Having an allowance of, say, $5000 to $10 000 for miscellaneous items in the budget will lessen financial stress for the owner builder, but most important is the need to try to stick to the budget, as the desire to save money is one of the most common reasons the individual is owner-building in the first place. The following list is some items that would fit into the miscellaneous budget:

- hoop iron (for columns)
- Bycol (additive for water content in mortar)
- brick ties
- damp course material.

11

Housing Energy Rating Scheme

Housing Energy Rating Scheme

In January 2003 the Australian Building Codes Board (ABCB) developed the Housing Energy Rating Scheme (HERS) as a national energy efficiency measure for homes across Australia. It was adopted by those Australian states and territories that did not already have an equivalent code in place. HERS is an index of a building's thermal performance, for example heating and cooling requirements, to determine how efficient a home is.

Under the scheme, houses are given a star rating between 0 and 10. Zero stars means the building does next to nothing to protect against the temperature outside. Ten stars will be very comfortable and use a lot less artificial heating or cooling. A five-star home is good, but is not outstanding. Since May 2011 HERS requires all new homes to have a minimum six-star rating.[1] There is still room to improve as technology develops and the means to quantify the thermal efficiency of homes progresses.

As you can imagine the large differences in temperature and rainfall between states and territories mean there are significant differences in heating and cooling needs. Hence there is a need to provide a number of climate zones that use different sets of parameters to rate thermal comfort level. Each home must comply with certain amendments, alterations or inclusions before it can be deemed to satisfy the thermal efficiency standards for the zone it is in. For this reason standards are set and assessed by each state, as described below.

Achieving sustainability

In 2009, the Council of Australian Governments, representing the Commonwealth and state and territory governments, announced that it would request the ABCB to increase the energy efficiency provisions in the 2010 edition of the Building Code of Australia (BCA). It was found that when Australian homes were measured against those in other countries with comparable climates that other countries performed better and had higher star ratings. That demonstrated that Australia was lagging in its efforts to reduce our carbon footprint.

By increasing our minimum star rating to six stars and including energy efficiency requirements for hot water and lighting (which had previously been assessed only in New South Wales) in new houses we can increase our energy efficiency to be comparable with other countries. The new six star energy efficiency rating takes effect on new homes built after May 2011.

Due to the importance of reducing our greenhouse gases a National Framework for Energy Efficiency (NFEE) was introduced and an agreement between all Australian governments was established to improve energy efficiency.[2]

There are several methods to satisfy our energy efficiency requirements. These include a building passing with a star rating issued using a thermal performance assessment. To satisfy the rating by this method you are not obliged to match all the acceptable construction provisions of the Deemed-to-Satisfy energy efficiency method, as the star rating shows that the building achieves the required overall thermal performance. A number of accredited software systems can be used to produce the required thermal calculation, including AccuRate, FirstRate5 and BERS Professional. These are described below.

The star rating provides a guide about the comparative energy performance of the home. The higher the number of stars, the more energy efficient the dwelling. A higher star rating would mean that the dwelling consumes less energy for heating and cooling, resulting in lower greenhouse gas emissions, and provides greater comfort.

What a sustainable home could mean to you is:

- improved comfort without using air-conditioning
- reduced running costs
- reduced greenhouse gas emissions
- reduced water consumption
- improved daytime lighting
- improved functionality
- improved resale
- improved marketability.[3]

The last two – improved resale and marketability – is catching on. Many people are choosing homes that are already fitted with sustainable energy improvements over others that do not have these features.

Software tools

Thermal simulation software such as AccuRate (formerly NatHERS), BERS Professional, and FirstRate5 are computer programs used to estimate the heating and cooling loads of a dwelling. Energy assessors enter the details of your plans or existing building into the software program. These programs, accredited by the Nationwide House Energy Rating Scheme (NatHERS), analyse the home's layout and orientation, as well as the construction of the roof, floor, walls and windows. This information is then matched with the local climate to calculate how much heating and cooling you will need to stay comfortable every day of the year.

These programs calculate thermal loads using standardised occupant behaviour for operating ventilation openings (windows and doors) and shading devices to ensure their effect on the loads is considered. The lower the load, the more the house can maintain comfortable conditions for the occupants without the need for air-conditioning or heating.

The calculated heating and cooling loads are simply the heat energy needed to be added or removed from the space and do not include the efficiency of the equipment used to heat and cool or the fuel type used – these are assessed in the Energy Index.[4]

NatHERS

The Nationwide Home Energy Rating Scheme (NatHERS) sets national standards for software used to rate the thermal performance capabilities of Australian homes. Homes are given a rating estimating the potential heating and cooling needs for the home. This allows fair comparison between different residential buildings. This tool calculates the heat energy gains and losses associated with the design of a building in a particular location and determine how much artificial heating and cooling may be required to maintain a comfortable temperature in the home.[5] NatHERS currently recognises AccuRate, FirstRate5 and BERS Professional as NatHERS-accredited thermal simulation software packages.

AccuRate

AccuRate is a thermal modelling rating software tool for residential buildings developed by CSIRO. It is an indicator of the heat needed to be added or removed to keep your home comfortable to live in. AccuRate assigns a star rating to a residential building based on its calculated annual heating and cooling energy requirements (CSIRO 2004). House energy ratings are a measure of the thermal efficiency of a dwelling, and the stars tell you how comfortable the home will be throughout the year.[6] The higher the star rating the less artificial heating and cooling is necessary. This software measures such things as the local climate and environment, the design of your home, and the layout of living areas and rooms. It looks at the construction of walls, roof, windows and floors, the orientation, and external shading of walls and windows.

BERS Professional

BERS (Building Energy Rating Scheme) is a NatHERS software tool based on CSIRO's calculation engine and incorporates many of the same improvements as AccuRate. BERS Professional has the added feature of a graphical data input process that lets designers draw house plans rather than typing in all the data. Much of the information about the building is selected from pictures displayed on the screen, making data entry quicker and easier.

BERS is most widely used in Queensland but can be used in all Australian climate zones. It is a powerful tool that is used to simulate and analyse the thermal performance of Australian houses in climates ranging from alpine to tropical and assign a star rating to a house within each particular climate type.[7]

NABERS

NABERS (the National Australian Built Environment Rating System) is a performance-based rating system for existing buildings and rates a building on the basis of its measured operational impacts on the environment. Initially developed by the former Federal Department of Environment and Heritage, NABERS is now being commercialised under licence by the NSW Department of Environment Climate Change and Water (DECCW).

NABERS HOME is an easy-to-use tool for comparing the energy and water use of an existing home to that of an average household. The web-based tool is available for anyone to use. The website also provides diagnostic tools, the Energy and Water Explorer, to provide personalised advice.

Because it focuses attention on the interaction between the occupants and the building, rather than the technical potential for that building, NABERS provides a realistic assessment of how a home is actually performing at a particular point in time as used by those occupants. The design of a home is only one factor, as performance is also greatly affected by choice of appliances and occupant behaviour.

NABERS is not a predictive tool. It complements, rather than replaces, other rating systems that focus on the design stage, such as HERS. It is designed to provide a simple indication of how well environmental impacts are managed, which gives building owners, managers or occupants the opportunity to reduce these environmental impacts.

NABERS extends beyond New South Wales' BASIX system (see below) to include commercial buildings and to measure ongoing performance of a building. It is also useful for councils, planning authorities and utility providers for assessing environmental impact and helping to establish incentives to reduce energy and water consumption and waste.

NABERS ratings for energy and water consumption apply to the following types of homes:

- detached dwellings
- townhouses
- duplexes
- villas
- terraces
- semi-detached dwellings.[8]

State provisions

New South Wales

BASIX is an initiative by the NSW Government that requires all new dwellings and all renovations costing more than $50 000 to be designed to achieve a 40% reduction in water consumption and 40% fewer greenhouse gas emissions compared to typical dwellings built in the past.

BASIX is administered by the Department of Planning and is a web-based planning tool that measures the potential performance of new homes against water, energy and thermal comfort indices. It offers a range of choices to gain approval:

- Insulation is mandatory, but you may use thicker insulation if it means you may pass your thermal comfort rating.
- Window glazing, eaves and window shading, awnings or shutters.
- Waste water for gardens.
- Use of light coloured roof materials.
- Landscaping that requires less water, using native plants and plants that are water-wise once established.

BASIX also encourages the use of these fixtures: gas or solar hot water systems, rainwater tanks for use in gardens, toilet and/or laundry; water-saving showerheads, taps and dual-flush toilets. If the rainwater tank is empty it will automatically revert to mains water, so you don't have to worry that you can't flush the toilet during dry periods. Showerheads and tap fittings should be a minimum AAA rating.

Each development application for a residential dwelling must be submitted with a BASIX Certificate. This is issued once an assessment has been satisfactorily completed, using the on-line tool on the BASIX website.

Since the introduction of BASIX in 2005 the Department of Planning NSW introduced AccuRate, BERS Professional and First Rate 5 Thermal Simulation Software as upgrades to the BASIX on-line tool. These contain a larger range of features in thermal assessment, making the assessment more flexible. They may include the use of:

- natural ventilation
- the cooling effects of ceiling fans
- heat flows in underfloor and roof spaces.

BASIX covers the building envelope thermal performance, but when determining compliance it also includes a wider range of household energy uses such as heating and cooling appliances, lighting and water heating. Currently the other states and territories do not use this method of assessment. However, the new six-star rating will require heating, cooling and lighting to be added into the equation. In common with many multiple issue tools, BASIX uses some existing tools such as NatHERS and appliance energy and water ratings as part of the assessment process.

To assess thermal performance compliance, you may use either the simulated heating and cooling loads predicted by NatHERS family software, or the building fabric must comply with DTS (deemed to satisfy) requirements (see below).

The simulation method provides more flexibility in design options. BASIX sets a maximum limit for both the cooling load alone and the total heating and cooling load. The simulation results must be less than these allowable maximums to achieve compliance. BASIX uses the HERS assessment to estimate greenhouse gas emissions, based on the thermal loads and the efficiency and type of heating and cooling appliances selected.[9]

A NatHERS certificate must be included with the BASIX certificate and it specifies the building materials used together with what type and thickness of the insulation is to be used in the new home.

Victoria

Sustainability Victoria uses FirstRate5 house energy software aligned with the national benchmark software, AccuRate and BERS Professional, to provide a calculation engine that includes:

- the ability to zone the house according to how each room will be used
- the ability to rate up to 10 stars
- the full range of AccuRate climate zones (11 in Victoria).

The FirstRate5 house energy rating software provides a simple and quick method to assess and improve the energy efficiency of house designs and completed homes.

The latest FirstRate5 has been the most popular HERS software in Victoria, although other NatHERS-family software can also be used. It is recognised as a house energy rating tool under the Nationwide House Energy Rating Scheme. Information relates to the proposed development and includes:

- types of materials to be used
- orientation
- insulation
- air leakage
- design features
- slab
- windows and glazing
- floor type
- zoning
- thermal mass
- width of eaves
- cross ventilation
- common walls.

If the rating is acceptable, an Energy Efficiency Rating Statement (EER) is issued.

In addition to achieving the star rating for the building materials, FirstRate5 also requires either a water tank that must supply water to all toilet cisterns, or a solar hot water system. The regulations require pressure reduction that will restrict water pressure to a building, and also flow restriction to showerheads and taps. Where there is reticulated gas, the solar hot water system must be gas-boosted.

The energy report will form part of the required documentation for building permit approval. As an alternative to using software packages, the Building Code of Australia (BCA) energy efficiency provisions can be used.

Importantly, a valid EER (Energy Efficiency Rating) is required for new house designs as well as for the sale of an existing house.[10]

Australian Capital Territory

The ACT's program is called ACT HERS, and is based on NatHERS. The minimum acceptable rating is six stars, assessed using FirstRate5. The program generates point scores based on design information and features such as:

- orientation
- insulation
- air leakage
- design features
- floor type

- zoning
- glazing
- thermal mass
- width of eaves
- cross ventilation
- common walls.

If the rating is acceptable, an Energy Efficiency Rating Statement (EER) is issued

The EER used in the ACT is confined to a rating of the thermal performance of the building shell because the rating scheme was originally developed from computer modelling of heat flows in building shells. It is designed to provide accurate and standardised information about building energy efficiency, excluding the hot water and lighting system, other fixed or movable appliances, and occupant requirements for temperature control. This model will more than likely be changed in line with the BCAB agreement to increase the minimum rating to six stars.

In line with changes to the BCA, the ACT moved to a six-star minimum for all residential homes built after May 2011.[11]

The deemed-to-satisfy energy rating method

Queensland, South Australia, Western Australia, Tasmania and the Northern Territory use the Building Code of Australia's energy efficiency provisions, which specifies the construction materials needed to achieve the home's energy efficiency rating. The BCA also allows for assessment through a 'deemed-to-satisfy' (DTS) method or through the use of computer software such as FirstRate5 and NatHERS. Software assessments give more flexibility in mixing materials and design options.

The BCA method is useful because it enables a building to be assessed relatively quickly and easily, but it provides less flexibility in improving the building design. It is fundamentally a check list of the BCA requirements, and either the proposed building complies with every requirement – and passes – or it does not.

Currently the BCA calls for a minimum energy rating for new houses of six stars, which is based solely on the design and construction materials used in a new home. Requirements vary depending upon the climate zone in which the building is located. The specifications have been progressively adopted in the Australian Capital Territory, Northern Territory, Queensland, South Australia, Tasmania, Victoria and Western Australia. New South Wales uses a different type of energy efficiency regulation called the building sustainability index (BASIX).

BCA energy efficiency provisions assess:

- the ability of the roof, walls and floor to resist heat transfer
- the resistance to heat flow and solar radiation of the glazing
- the sealing of the house

- the provision of air movement for free cooling, in terms of openings and breeze paths
- the insulation and sealing of air-conditioning ductwork and hot water piping.

The provisions are planned around the following eight climate zones:

1. Hot humid summer, warm winter.
2. Warm humid summer, mild winter.
3. Hot dry summer, warm winter.
4. Hot dry summer, cool winter.
5. Warm temperate.
6. Mild temperate.
7. Cool temperate.
8. Alpine.

Requirements for each zone depend on a number of climate characteristics.

Under DTS assessment an applicant can use any materials, components, design factors or construction methods that comply with the relevant performance requirement. The key to the performance-based DTS is that there is no requirement to adopt any particular material, component, design factor or construction method, and this provides you with the flexibility to choose innovative architectural design.

The BCA still caters for those who require specific guidance or who wish to continue to use traditional building methods. The DTS provisions of the BCA continue to provide detailed prescriptive methods for establishing compliance with the performance requirements.[12]

Approval may still be issued to a design if it differs in whole or in part from DTS provisions described in the BCA if it can be demonstrated that it complies with the relevant performance requirement. For more information or alternative solutions, visit <www.bcab.gov.au>.

Some states have slight variations in the DTS requirements as shown below. In line with changes to the BCA, all states moved to a six-star minimum for all residential homes built after May 2011. This requires stringent assessment and will most likely be assessed inclusive of heating and cooling treatments and hot water supply.

Queensland

Queensland's Sustainable Housing initiative has proposed additional requirements beyond the BCA. These require all new houses to have:

- greenhouse-efficient hot water systems
- energy efficient lighting

- AAA-rated showerheads
- dual-flush toilets
- water pressure-limiting devices
- rainwater tanks (not required by all councils).

In units and apartments, the standard will require:

- energy-efficient lighting
- AAA-rated showerheads
- dual-flush toilets.

The standard also affects bathroom renovations, requiring:

- AAA-rated showerheads
- dual-flush toilets.[13]

The Queensland DTS model may give you an additional star or half a star for providing an outdoor living area, subject to a defined criteria, you can find this on the Building Code Australia website.

South Australia

South Australia has no plans to develop its own standard, so will be sticking with the BCA. However, the SA Housing Code also regulates:

- sealing capacity
- air movement
- hot water services.

In addition, external glazing and/or shading is required, such as shutters or awnings. Plumbed rainwater tank tanks must be 1000 litres and plumbed into a toilet or laundry. Roofs, external walls and suspended floors must achieve a minimum total thermal resistance.[14]

Western Australia

Dual-flush toilets have been mandatory for a number of years through the state's plumbing regulations.

Tasmania

Tasmania has no plans to move away from the BCA. For assessment, Tasmania uses only the DTS method and does not provide the option of using a computer-generated model to assess the design.

DTS allows the builder to submit a design that meets a set of minimum criteria.[15]

Northern Territory

The Northern Territory government is contemplating the adoption of the BCA 2006 five-star energy efficiency requirements. If adopted, the NT Government is undecided as to what method of assessment they will use but currently the preference is for DTS BCA provisions.

The NT Government's Building Sustainability Services (BSS), Department of Planning and Infrastructure, provides technical policy advice and services, benchmarking and reporting tools, and oversees projects to promote ecologically sustainable NT Government infrastructure. BSS also provide useful information on sustainability for organisations, schools, households and the public in general.[16]

Other tools and organisations

The **Australian Building Greenhouse Rating Scheme** is used to assess the energy and greenhouse efficiency of buildings, and the NABERS office water scheme is used to assess water usage. A full range of measured operational impacts including assessing waste and indoor air quality are being developed to add to the environmental impacts considered.

The **Australian Sustainable Built Environment Council** (ASBEC) is a peak body of key organisations committed to a sustainable built environment in Australia. ASBEC currently has 29 members comprising mostly industry and professional associations with members who are involved in the planning, design, delivery and operation of our built environment, and those concerned with the impact of the built environment upon society.

The **Association of Building Sustainability Assessors** (ABSA) is a not-for-profit incorporated association representing building and design professionals who specialise in assessing the environmental impact of buildings.

ABSA accredits assessors to conduct assessments required by certain building and planning regulations, such as the Building Code of Australia and the NSW Building Sustainability Index (BASIX) Thermal Comfort Simulation Method.

Water Efficiency Labelling and Standards (WELS) Scheme can assist purchasers of household water-using products to compare the relative water efficiency of the available models.

Smart Approved Water Mark is Australia's water-saving labelling program for products and services that are helping to reduce outdoor water use.

Your Home is a suite of consumer and technical guide materials and tools developed to encourage the design, construction or renovation of homes to be comfortable, healthy and more environmentally sustainable.

Sustainable Housing in Central Australia is a guide to efficient use of energy and water for buyers, builders and renovators in Central Australia.

Green Smart is an industry-driven initiative to encourage a mainstream application of environmentally responsible principles to housing design and construction.

The Master Builders Association (MBA) provides a training program called **Green Living**. The Territory Construction Association is the Northern Territory contact for the MBA.[17] Websites for New South Wales and Victoria are listed below.

Useful information

Please use these sites – they are the experts and will help you get the most out of this chapter. Our experience in contract housing is basically for New South Wales, and now the rest of Australia is catching up to being more sustainable by using less fossil fuels and saving water.

AccuRate: <www.hearne.com.au/products/accurate>
BCA (Australian Building Codes Board): <www.abcb.gov.au>
BERS Professional: <www.solarlogic.com.au/bers-pro/details>
Council of Australian Governments: <www.coag.gov.au>
FirstRate5: <www.sustainability.vic.gov.au/www/html/1491-energy-rating-with-firstrate.asp>
NABERS: <www.nabers.com.au>
NatHERS (National House Energy Rating Scheme): <www.nathers.gov.au>

Useful contacts

Assessors: national

See also state listings.
ABSA (Association of Building Sustainability Assessors): <www.absa.net.au>

Australian Capital Territory

ACT Planning and Land Authority: <www.actpla.act.gov.au/topics/design_build/siting/energy_ratings>

Assessors

ACT Planning and Land Authority: <www.actpla.act.gov.au/topics/design_build/siting/energy_ratings>

New South Wales

Department of Planning: <www.basix.nsw.gov.au/information/index.jsp>

Master Builders Association: <www.mbansw.asn.au>
New South Wales Department of Planning and Local Government: <www.planning.nsw.gov.au>

Northern Territory

Northern Territory Lands Group: <www.nt.gov.au/lands/building/energy/index.shtml>

Queensland

Building Codes Queensland: <www.dip.qld.gov.au/plumbing-building/index.php>

South Australia

South Australian Department of Planning: <www.planning.sa.gov.au/go/building>

Assessors

Planning SA: <www.planning.sa.gov.au>

Tasmania

Workplace Standards Tasmania: <www.wst.tas.gov.au/industries/building>

Victoria

Master Builders Association: <www.mbav.com.au>
Sustainability Victoria: <www.sustainability.vic.gov.au/www/html/1491-energy-rating-with-firstrate.asp>
Victorian Building Commission: <www.buildingcommission.com.au>

Assessors

Sustainability Victoria: <www.sustainability.vic.gov.au/www/html/1796-accredited-energy-raters.asp>

Western Australia

Office of Energy: <www.clean.energy.wa.gov.au/pages/her_tools.asp>

References

Australian Greenhouse Office (2004) *Australian Residential Building Sector Greenhouse Gas Emissions 1990–2010.* Department of Environment and Heritage, Canberra.

Australian Greenhouse Office (2005) *Your Home Technical Manual, Rating Tools.* Commonwealth of Australia, Canberra.

Climate Action Network Australia (2005) Inquiry into Energy Efficiency. Submission to Productivity Commission Review of Energy Efficiency.
CSIRO (2004) *AccuRate Manual. The CSIRO Home Energy Saving Handbook.* CSIRO Publishing, Melbourne.
Gregory KE, Moghtaderi B, Sugo H, Page A (2007) Effect of thermal mass on the thermal performance of various Australian residential construction systems. *Energy and Buildings*, 2 April.

12
Bushfire protection

Standards for bushfire-prone areas

The devastating Black Saturday bushfires in Victoria of February 2009 prompted a revision of the rules about bushfires. The fires reached temperatures above 1200°C and wind speeds were more than 120 km/h, and in those extreme conditions the 'stay and defend' strategy proved fatal for many home owners. These figures prompted a rethink about how the fire danger warnings are given so that people have the information to decide whether to evacuate, and they highlight the need to design residential properties to withstand extreme conditions.

In response to the Black Saturday fires the authorities throughout Australia introduced a new standard in bushfire management, with the implementation in March 2009 of the AS3959-2009 for the construction of buildings in bushfire prone areas. This legislation covers all new dwellings, including alterations and additions as well as out-buildings, pergolas and the like, to ensure that homes are built to a higher degree of safety so they are better equipped to be protected from fire attack.

The expectation is that these design principles, when properly implemented, will not only protect your family and yourself, your home and possessions, but also the members of the fire-fighting services. This new legislation will ensure that residential and other developments are not unduly exposed to risk from high-intensity bushfires.[1] In addition they will reduce costs to the community in cleaning up work, insurance claims and government support, not to mention helping the environment.

These design principles cannot guarantee that your home will withstand a bushfire. This chapter is a guide to what the owner builder needs to address when building in a bushfire prone area and is not a definitive guide on bushfire protection.

Assessing your requirements

To find out if you are in a bushfire zone, maps are available for inspection at your local council. You can purchase an up-to-date 149 certificate from your local council in NSW. This is a certificate issued in accordance with the Environmental Planning & Assessment Act 1979 and contains information on how a property can be used and the restrictions on development. You need an up-to-date 149 certificate as they changed due to the implementation of new legislation. Other states may have something similar; either way, the Rural Fire Service will help you to determine if you are building within a bushfire asset zone.

With ever-increasing land subdivisions moving further into the Australian bush and the desire to maintain natural flora and fauna in new land developments and estates, you may find the block of land you like is in a bushfire asset zone. If that is the case, you will be required to include additional safety features in the design and construction of your home.

If you are in a bushfire asset zone the developer will advise you of what your responsibilities are in the form of design control guidelines over and above the standard design control guidelines for that estate. What will have to be included to protect you and your home from a bushfire depends on how close your block is to the bushfire asset zone.

Apart from new estates and subdivisions, if you live in a vegetated area you need to determine if your land is considered a bushfire asset zone. It is vital that you do an assessment before the design stage of your home so you can incorporate the building materials and features required into the construction and not waste time and money adding these requirements at a later date. You can make the assessment yourself or contact the relevant agency in your state from the list provided at the end of the chapter.

In addition to these websites and to help you assess your Bushfire Attack Level (BAL), <www.buildingcommission.com.au> has a nine-page pdf that is easy to understand and will give you an overall guide to the steps and processes required to determine your bushfire attack level.

If you wish to discuss how to do determine your BAL you should contact a bushfire assessment consultant in your state. You can find one through a web search using these key words: 3959-2009 bushfire planning. There will be a number of bushfire assessors in your state to give you advice. You can also contact your rural fire service or local council for more information.

Depending on how close you want to build to vegetation and the type of vegetation on your land, you could also require an asset protection zone. An asset protection zone is a clearing or buffer between your home and the vegetation on your

Table 12.1 Bushfire attack levels and associated construction requirements[2]

Bushfire attack level	Construction requirements
BAL-LOW	Minimal attack from radiant heat and flame due to the distance of the site from the vegetation, although some attack by burning debris is possible.
BAL-12.5	Attack by burning debris is significant with low levels of radiant heat (not greater than 12.5 kW/m^2). Radiant heat is unlikely to threaten building elements (i.e. unscreened glass). Specific construction requirements for ember protection and accumulation of debris are warranted (Level 1 construction standards).
BAL-19	Attack by burning debris is significant with an increased radiant heat levels (not greater than 19 kW/m^2) threatening some building elements. Specific construction requirements for protection against embers and radiant heat are warranted (Level 2 construction standards).
BAL-29	Attack by burning debris is significant and radiant heat levels (not greater than 29 kW/m^2) can threaten building integrity. Specific construction requirements for protection against embers and high radiant heat are warranted. Some flame contact is possible.
BAL-40	Increased attack from burning debris with significant radiant heat and the potential for flame contact. The extreme radiant heat and potential flame contact could threaten building integrity. Buildings must be designed and constructed in a manner that can withstand the extreme heat and potential flame contact.
BAL-FZ (FLAME ZONE ALTERNATIVE SOLUTION REQUIRED)	Radiant heat levels will exceed 40 kW/m^2. Radiant heat levels and flame contact are likely to significantly threaten building integrity and result in significant risk to residents who are unlikely to be adequately protected. The flame zone is outside the scope of the BCA and the NSW Rural Fire Service may recommend protection measures where the applicant does not provide an adequate performance solution. Other measures such as drenching systems and radiant heat barriers may also be required.

land. The width of the asset protection zone is determined by the slope and distance to the vegetation. If you do require an asset protection zone and or you would like to undertake fire hazard reduction you must apply for permission through your rural fire service. Depending on which state you live in, a good place to start is the websites listed at the end of this chapter. The assessment is a free service.

Once you have recognised that you are in a bushfire zone you need to know what your responsibilities are. Using the site classification procedure, determine the bushfire hazard attack level in your area. Each level requires different measures to render a home less susceptible to fire damage, as summarised in Table 12.1.

The categories of attack are determined by:

1. The type of vegetation
2. How close your building is to vegetation
3. The effective slope (fire runs more readily and with greater speed uphill)
4. The fire danger index applicable to your region.[3]

Designing for fire protection

Every summer we are met with the possibility of bushfires. The risk is high, so it is important to be prepared. Radiant heat and ember attack are as serious a risk to properties in bushfire prone areas as direct fire attack. We must accept that fire is a part of our unique ecosystem, and we need to plan and prepare to reduce the severity and impact of fire when it does strike.

Planning for Bush Fire Protection 2006 <www.rfs.nsw.gov.au> details six key bushfire protection measures that can be incorporated into a development to improve protection against bushfire attack. No single protection measure provides sufficient protection from the impacts of bushfire. However, an acceptable level of protection can be achieved by providing a combination of these bushfire protection measures to a development. Any application to develop in a bushfire prone area must consider the six bushfire protection measures. The measures include:

1. Asset protection zones (fuel reduced areas)
2. Access arrangements
3. Building construction and design
4. Water supply and utilities
5. Landscaping
6. Emergency management arrangements.

For more information about each of protection measures, go to the Bush Fire Protection Measures page given at the end of this chapter.

In a paper published in 1945, G.J. Barrow described how incorporating certain details into your home design and its finishes would considerably reduce the risk of a home succumbing to bushfires.

In a detailed study of houses destroyed by fire in the (then) urban fringe Melbourne suburb of Beaumaris, 14 January 1944, he determined that the resistance of houses to external fire hazards is determined more by the details of construction than by the materials used in the walls. Although the damage was caused primarily by the external fire, practically all the houses ignited inside, that is in the roof space, in rooms or under the floors, as the result of the entry of flame, sparks and embers through openings such as ventilators, eaves and windows. Sealing or screening such openings with fine wire mesh greatly reduced the risk of damage, and the conclusion was drawn that, from the point of view of resistance to external fires, a house should be as air-tight as practicable and that any openings which cannot be eliminated should be screened.[4]

There are several steps to help you design a home built to resist bushfire attack. The site where the following advice was sourced is no longer current, but the information remains relevant. Check your state requirements before you build as specifications can change.

1. Try to build on the flattest part of your land. Avoid steep slopes, especially if they are north- or west-facing. Avoid any hazardous vegetation such as overhanging branches from trees and thick bushes.
2. If you are 30 m or more from the roadside, site your home where you can provide an easy-to-use track so it is safe for fire trucks to enter, leave, and move around your home.
3. Provide a water supply that is just for bushfire use, even if you have to purchase the water. Have the right connections for fire fighters to connect their hoses, and provide a diesel or petrol-fuelled generator for pumping the water.
4. Build your home using techniques and materials that make it more resistant to bushfires. There is a list of providers of building materials that are designed for bushfire prone land.
5. Flooring should be one of the following: a concrete slab on the ground, a suspended concrete floor, or a timber or steel-framed floor where the underneath of the bearer is more than 600 mm above ground level. In this case the space under the floor should be protected by a non-combustible sheet material such as fibre cement.
6. Windows should be screened with corrosion-resistant steel, bronze or aluminium mesh with a maximum aperture size of 1.8 mm.
7. External doors should be fitted with weather strips. Screen doors should be tightly fitting and use corrosion-resistant steel, bronze or aluminium mesh with a maximum aperture size of 1.8 mm.
8. Vents and weepholes should be protected with corrosion-resistant steel, bronze or aluminium mesh with a maximum aperture size of 1.8 mm.
9. Roof coverings should be made of metal sheets or fibre-reinforced cement. Do not use timber shakes or shingles. All gaps under corrugations in the sheet roofing should be sealed or protected in one of the following ways:
   - Full sarking directly under the roof
   - Covering gaps with a corrosion-resistant steel or bronze mesh with a maximum aperture size of 1.8 mm
   - Using a profiled metal sheet
   - Using a neoprene seal or compressed mineral wool.
10. Cappings on sheet roofing should be pre-formed or the gaps between the capping and the sheeting should be sealed or protected.
11. Tiled roofs should be fully sarked with a flammability index of no more than 5, including the ridge, and the sarking should be located directly beneath the tiling battens.
12. The roof–wall junction should be sealed with a fascia and eaves lining, or the gaps between the rafters at the line of the wall should be sealed with a non-combustible material.

13. Penetrations through the roof cladding for vent pipes and the like should be sealed with a non-combustible collar or fire-retardant sealant.
14. Skylights and other shafts through the roof space should be sealed with a non-combustible sleeve or lining. Thermoplastic sheet in a metal frame may be used, provided that the diffuser at ceiling level is made of wired or toughened glass in a metal frame. Openings in ventilated skylights should be covered with corrosion-resistant steel or bronze mesh with a maximum aperture size of 1.8 mm.
15. All components of roof ventilators, including rotary ventilators, should be made of non-combustible material and have their openings protected by corrosion-resistant steel or bronze mesh with a maximum aperture size of 1.8 mm.
16. All openings into roof-mounted evaporated air cooling units should be protected by corrosion-resistant steel or bronze mesh with a maximum aperture size of 1.8 mm.
17. Gutters and downpipes materials or devices used to stop leaves collecting in gutters should have a flammability index of no more than 5.
18. Service pipes (water and gas) piping for water and gas supplies should be buried at least 300 mm below the finished ground level, or be made of metal.
19. Verandahs and decks should be made of concrete, and the posts, columns or supporting walls must be adequately protected. If you want a timber deck, the gaps between the boards should be no smaller than 5 mm. So that a fire under the deck can be put out, the perimeter of the deck should not be enclosed.[5]

These measures are by no means exhaustive, and the owner builder must research his/her Bushfire Attack Level requirements. **The measures contained in this chapter cannot promise that your home will survive a bushfire event** as there are other factors such as fire behaviour and weather conditions on the day and the degree of vegetation management.

An important measure is reducing fuels such as dried leaf materials, sticks, twigs, bark, grass, and so on lying around your home. Landscape your garden using low, bushfire-retardant, plants: you can find a list of these in the fact sheets in your state or territory.

References

Barrow GJ (1945) A survey of houses affected in the Beaumaris fire, January 14, 1944. *Journal of the Council for Scientific and Industrial Research* **18(1)**, February.

Bushfire Protection Measures: <www.rfs.nsw.gov.au/dsp_content.cfm?cat_id=1058>

Fact File: Bushfire Construction Requirements 2010. <www.mornpen.vic.gov.au 22/09/2010>

Ramsay C and Lisle R (2003) *Landscape and Building Design in Bushfire Prone Areas*. CSIRO Publishing, Melbourne.

www.cfa.vic.gov.au/firesafety/buildingandregulations/building-home.htm

Useful websites and contacts

Bushfire risk assessment

National

www.buildingcommission.com.au/resources/documents/Building_Commission-BAL_Guide_(3).pdf

New South Wales

www.rfs.nsw.gov.au or http://hia.com.au

Queensland

www.ruralfire.qld.gov.au

South Australia

www.planning.sa.gov.au/building_policy and SA Country Fire Service Development Assessment unit (bushfire Protection) <www.cfs.org.au>

Victoria

www.buildingcommission.vic.gov.au, your own private building surveyor, or from the Shire's Statutory Building Team

Western Australia

www.fesa.wa.gov.au Fire and Emergency Services Authority and Western Australia Planning Commission <www.planning.wa.gov.au>

Useful contacts and publications

Publications from the Rural Fire Service NSW, 1 October 2009: <www.rfs.nsw.gov.au>

- Prepare. Act. Survive. Publications.
- Fire Danger Ratings. Publications.
- Bush Fire Survival Plan. Fact sheet.
- Leaving Early. Fact sheet.
- Bush Fire Preparation. Fact sheet.
- Bush Fire Alerts. Fact sheet.

www.cfa.vic.gov.au/firesafety/buildingandregulations/building-home.htm

Technical advice

CSIRO Forestry and Forest products: <www.ffp.csiro.au/nfm/fbm>
James Hardie: 13 11 03 <www.jameshardie.com.au>
Hebel Australia: 1300 369 448 <www.hebelaustralia.com.au>
Plywood Association of Australia
  3 Dunlop Street
  Newstead Qld 4006
  Tel: (07) 3854 1228; Fax: (07) 3252 4769
Think Brick Australia: (02) 9629 4922 <www.thinkbrick.com.au>

Timber Advisory Services

New South Wales

Timber Development Association NSW Ltd
13-29 Nichols Street
Surry Hills NSW 2011
Tel: (02) 9360 3088 Fax: (02) 9360 3464

Queensland

500 Brunswick Street, Fortitude Valley Qld 4006
Tel (07) 3254 1989 Fax: (07) 3254 1964

South Australia

Timber Development Association of SA
113 Anzac Highway, Ashford, SA 5035
Tel: (08) 8297 0044 Fax: (08) 8297 2772

Tasmania

Tasmanian Timber Promotion Board
38 Montpelier Retreat, Battery Point TAS 7004
Tel: (03) 6224 1033 Fax: (03) 6224 1030

Victoria

Timber Promotion Council
320 Russell Street, Melbourne Vic 3000
Tel: (03) 9665 9255 Fax: (03) 9665 9266

Western Australia

Timber Advisory Centre
Cnr Salvador Road & Harborne Street,
Wembly WA 6008
Tel: (08) 9380 4411 Fax: (08) 9380 4477

13

Knockdown rebuilds

There are many reasons why your home may no longer meet your needs. For example, it could be that your existing home doesn't provide you and your family the space and comfort you long for; one or both your parents may have to move in with you; your extended family may need to be accommodated; or perhaps simply you have just outgrown and are tired of your old home. In these circumstanves you will want to add on, move, or knock down and rebuild. Each of these options comes with its own pros and cons.

Renovation

The cost of renovating or adding on can be significantly more than the cost of a knockdown rebuild. You need to also ask if your existing home is in a condition that is suitable for an extension. Will it be easy to design a functional home around your existing home? Consider the upheaval of adding on and the inconvenience of living out of one room at a time during the remodelling. This disruption can take its toll on your family.

Access to the building area is much harder when you are adding on rather than rebuilding, and you may have to employ cranes to get your materials into place. You need to consider whether you can match materials to your existing home. Accurate matching can add an enormous amount of money to your extension and you still may not get what you envisaged.

Then there is the danger that every step you take uncovers something else that needs fixing. This can easily happen to any home renovation if you aren't prepared for all the possibilities that a renovation entails. If you are not careful you will have to keep throwing money into it to fix problems.

In addition to this, if your renovations are extensive you will more than likely have to comply with sustainability requirements such as installing a rainwater tank and energy saving devices like insulation and low voltage lighting to reduce your carbon footprint.

There is also a possibility that there could be other requirements, such as heritage overlays, so it may not simply be a matter of building what you envisage, even if you can afford it.

To give you an idea on your outlays, adding on or renovating generally costs twice as much per square metre as building a new home. Renovations can blow the budget, not to mention present you with unforeseen construction problems, restrictions and compromises.

After all this, even if you can extend or add on slightly more cheaply, the rest of the house is still old. Building a new home enables you to start with a blank sheet.

Building on a new site

The next option is to consider is purchasing land in a greenfield site. However, increasing urban sprawl means you will have to go further and further from your existing suburb and way of life if you are to purchase land to build new. Knockdown rebuild is possibly the only way to build a new home in an established suburb.

Purchasing a block of land in a new estate or subdivision and building a home once cost about the same. Now, however, with developers' contributions, infrastructure and the like, the cost of purchasing a block of land far outweighs the cost of building a new home. You also have to sell your old home and pay agents' fees, legal costs and stamp duty levies. If you find somewhere you think you could call home, you may not be able to be an owner builder or even build with your preferred builder, as some new estates have their own preferred builders who you must choose from.

New greenfield estates can have other restrictions too. In most, if you have a truck of more than 2 tonne you will not be able to park it, so check the land contract – some have no limits. Many estates have additional design control guidelines that must be adhered to, over and above government sustainability requirements; these may specify building materials and construction methods, or set minimum floor areas. You may have other additional charges; for example if you purchase a corner block most estates require the fencing to be masonry with lapped and capped fencing or some other type of insert for privacy.

A renovation black hole

I recently bought a fabulous lounge suite. It looked great in the showroom and it looked about the same size as my old one, so I eagerly awaited its arrival. (It took 5 months to make.) The day it arrived I had to go to work, and it wasn't till I got home that I realised to my dismay that it was much too big for my 1950s 11 sq home. The lounge cost $12 000 and I couldn't send it back.

The only option I saw was to engage a carpenter to extend my lounge room. Luckily for me I did have the space over the front porch, and I worked for a builder who could do the work for me. The supervisor came out and said it could be done for around $2000, so I thought great – when can you do it? He said he would send one of the boys over as he didn't have time to do a small job like that. It took a few phone calls to get one of the boys to come out and he had a look in the roof to make sure the roof would not collapse when he took the front wall out. He gave me a more realistic quote of $8000, which of course I wasn't prepared for. Adding to the expense, I had chosen James Hardie Matrix to finish off the exterior wall and this cladding is labour-intensive to install. I had to call a halt to this as the builder didn't have time to do the job – too small, and I didn't have the money.

In the end a friend who was a builder started the job, for which I paid a flat rate per day, and I saved a lot on materials by purchasing them all through work. Then when that money ran out I had to leave it for a few months until my brother had time to do a bit, then a few more months when he could spare the time to finish it structurally. Luckily for me I had some under-eave panelling that I had bought on eBay to use as under flooring when I did my tiling, so I was able to use that instead of the James Hardie Matrix.

The expenses kept mounting: I also had asbestos removal and tipping fees. I could not match the tiles I had laid only a few months pre-lounge – I had gone to eBay and purchased 32 m^2 of polished porcelain tiles at a bargain price from someone who had over-ordered on his job. I couldn't match these anywhere so I went for something close to the colour. The new tiles were the big 500 mm by 500 mm and my tile cutter wasn't big enough to hold them, so I had to take them to a tile cutter. Now I have tiles in two different sizes and colours in my living room. The panelling worked well but it did not match the old cladding we did in the 1970s. All this work did give us the opportunity to install wall insulation into the new lounge room, which is really good.

My only option now is to take the rest of the old panelling off and paint the whole front of the house. It still isn't finished. I have to do all the sanding and painting, plus I had to buy a new front door to replace the outdated panel door. This meant another tri lock, because the new front door was a lot more modern (I bought it on eBay really cheap).

The moral of this story is to take a tape measure with you when you purchase a lounge suite! Another moral is that it shows how easy it is to get caught out with renovations, even the planned ones.

Best of both worlds?

The knockdown rebuild is a very popular way to have the best of both worlds. You already have the block of land on which to build, and you will remain in the area that you have lived in and are familiar with, and with the same relationship to schools, work, bus and train services, shopping centres, friends and neighbours. At the same time you will have a new, smarter, more energy-efficient home with all the latest design and lifestyle benefits.

Research before you start

Before you start it is generally a good idea to find out recent sales prices in your area so you do not over-capitalise on the construction of your new home as you will want to receive the best return on your investment.

If you have decided to go down the knockdown rebuild path, it is just as important to find out every possibility that could affect your plans as if you were building in a new estate. At the top of the list you should find out if you live in a heritage-listed or conservation house, a 1:100 year flood zone, under an aeroplane flight path or near a railway line; if you in a cyclone or high wind area, a mine subsidence area, bushfire zone or near a lake or the beach.

If you live in an area where your home is heritage listed you may not have the choice to knock down and rebuild, or indeed to renovate, without significant compromise. Sometimes your only option is to move to build new after all. You should seek council advice as to what your options are. If you live in an area affected by any of the other considerations mentioned you need to speak to your council to find out what, if any, design elements would need to be included. You could enlist the help of a qualified private certifier who is experienced in building with your council and building in your area.

You can also get ideas of what has been permissible simply by driving around your area to see what developments are under way – ask the owners how they went about their approval process.

The knockdown

Now you have established that you can knockdown rebuild, you need to retain the services of a demolition expert. As with any service, you should shop around to get the best value your money can buy. Not all demolishers will offer the same inclusions, so it pays to shop around and be aware that the cheapest price may not be the best and it may not include everything that is required to make your site ready for construction.

You will need to start by asking to see the demolisher's licence. Check to see that the licence and the contractor's name match. You may also do a check with Office of Fair Trading and ASIC to see if they have had any disputes raised from previous jobs.

Check how much insurance liability they have and if their insurance is up to date. They should have a minimum of $10 000 000 in case there are any complications.

Ensure they will remove everything that needs to go from the building envelope so that the construction site is clean of all building materials, vegetation, driveways, retaining walls and anything else that may be in the building zone.

You will have to contact your utilities providers to disconnect your services, power, gas, water and sewer connections, and ask if they will cap them off so you can reconnect later.

When these services and old foundations are all removed it will create large holes in your construction site and undermine the existing ground. You may have to import soil to fill the holes and compact it, along with increasing the strength of the new slab for your home.

You will also have to be especially cautious of asbestos and any other potentially dangerous items. Your demolisher must have the relevant expertise for the removal and disposal of it.

If you have a tree or vegetation in the building envelope, apply to council for its removal with the demolition application. This will save money in DA lodgement fees.

You will need to give the demolisher all the information about your new project so they know exactly what needs to be done to prepare your site. Make sure they provide council and statutory costs in the quote. They will need to provide a security fence around the site as well as a sediment barrier to stop any debris from going into the stormwater drain, and you may have to pay a kerb and gutter bond in case they damage it.

Ask the demolishers if you can recycle materials from your home such as hardwood flooring, windows, doors, kitchen appliances, the kitchen, and anything you feel is of value. This must be organised before work commences. You can resell to renovators, advertising these materials for sale, or you can have a demolition sale. As long as you are covered by insurance you can get the purchaser to bring their own tools to dismantle their purchase. This will have to be organised for another day and at a time that is convenient to both of you. Just get half the money upfront in case they change their mind, or you will be left holding the baby. You could also advertise the whole house for removal to another site. This is called a humpy. There are experienced removalists who will cut the house in half and load it on a flat-bed truck and move to its new home. You will also be helping the environment and making some extra money for your project.

Planning the rebuild

When you are budgeting, don't forget to allow money for retaining walls, garden beds, new plants, paths, driveway, clothesline, floor coverings, window furnishings, fly screens, security screens and light fittings, as with a new home anywhere.

Additional expenses in a knockdown rebuild include storage of items if they won't fit in your rental accommodation, and extra money for hiring the removal van for two moves – from your existing home to the rental accommodation, and then when your home is ready, for your move in to your brand-new home. You will have to pay to have your utilities reconnected, and pay the associated bonds required for each of them.

The next thing you need to do is organise where you will live while you build your new home. Make sure you have a flexible lease in case it takes longer than you planned to build. Try to find something close to your site so you can keep an eye on it and its progress.

There is a lot to do! So get started write a list and then tick everything off as you go along. this is an exciting and probably daunting time but one we am sure you will be glad you did. You may like to design a spreadsheet and enter everything you need to do in sequence so you don't forget anything.

14

Contracts

Signing contracts

A contract is an agreement between two or more parties. It spells out the terms of the agreement and is a legally binding document. As an owner builder you will be entering into several contracts as the need arises during the construction of your home.

All building trade contractors who perform work on your site that costs over a certain amount must have a contract in writing outlining the work that is to be performed, the materials and costs. Both parties must sign, and each retain a copy. The cut-off amount varies: it is $1000 in New South Wales, in Victoriait is $5000, in Western Australia it is $7500, in Queensland it is $3300, in South Australia it is $5000 and in Tasmania it is $5000.

A contract can be fairly simple and written in plain English. The New South Wales Fair Trading website has free downloads of a plain-English contract for home building contract work costing under $25 000 and one for home building contract work over $25 000. Each state has an equivalent contract available, but this one is easy to read and concise; in any event from 1 January 2011 all Australian consumers wherever they live have been protected under Australian consumer law and there is one set of rules for everyone.

The standard New South Wales contracts include a guide in the form of a checklist: you or your contractor, whoever is supplying the contract, tick the boxes as you go. When you have all the boxes ticked and you are sure nothing is missing, the contract is ready to sign. This means you will have all the necessary statutory warrantees, insurances and licences, costings for materials, plans or drawings, and a description of the works to be carried out and anything else that may be required before you pay a deposit and any work commences.

The signing of your building contract does not need to be scary. These days most contracts are self-explanatory and written for the layperson, not like the old days where you needed to know the legal jargon to decipher what it said and engage a solicitor to check it over – and who would charge you a few hundred dollars to do so.

Our advice is take the contract home to read at your own pace. Don't feel obligated to sign anything until you are completely satisfied and understand the jargon of building and construction processes.

If you do not understand the contract you should seek advice from a solicitor, a qualified architect, or the Home Building Advisory Service. There is also a Consumer Building Guide you can download from the internet <www.fairtrading.nsw.gov.au/pdfs/About_us/Publications/ft246.pdf> that explains all the processes that are involved from start to finish and what to expect.

There is also a cooling-off period. If for some reason the contractor does not provide you with a signed copy of the contract within 5 working days you may rescind the contract.

It is very easy to lose track of changes that can and do get made on the spot. This can cause misunderstandings and frustration and can lead to disputes if one of you forgets that you both agreed to the variation. Therefore, if during construction you change something or your inclusions change, you add something or take something out, it is vital that these changes are written as a variation that includes a description of the variation along with costs for labour and materials. This must be signed and acknowledged by both parties before the variation is implemented.

Types of contract

There are a few different types of contracts.

Fixed price or lump sum contracts purport to state the total amount for the work, although it isn't uncommon for the final cost to increase as builders may from time to time have not estimated something into the contract due to unforeseen circumstances. Very often builders may not have included enough in the contract for the piering, as this is usually a provisional sum and determined on the day of the pour by an independent engineer. If your contract does not have at least 50 lineal metres of piering costed you will more than likely be up for additional costs as most homes require more piering than that. Nearly all builders put in the minimal cost to make their quote look more attractive than if they had put in a more realistic piering allowance. At the time of writing the cost for piering is usually about $88.00 per lineal metre. The extra piering will be raised as a variation to the contract and is one of many out-of-pocket expenses that could crop up during your construction. However, you must be diligent and ask questions, because if the builder has made a mistake then it should be rectified by him at no expense to you.

A cost-plus contract can be cost effective, as you pay only for the materials used plus an hourly rate. This is how we paid for our extension. You should get an estimate in writing as to time frame, materials and hourly rate so you have an idea of what to expect. Ask questions if you do not understand how they worked it out. If this is the way you want to go you should request all receipts before you pay for goods.

The design-and-construct contract is another type of contract where by the home owner provides a basic brief to the contracting organisation, who then supplies the designs and builds your home.

You may also ask about warranties or guarantees and want them written into your contract. Statutory conditions and warranties are implied by statutes or common law and require traders and manufacturers to ensure that every product provided is fit for the purpose for which it is supplied.

- The service must be carried out with due care and skill and any materials supplied will be reasonably fit for the purpose for which they are supplied.
- The service must be carried out in accordance with all relevant laws in an appropriate and skilful way.
- It must be in accordance with plans and specifications.

So even if a warranty is not written into your contract it is implied that you are covered for defects and bad workmanship because you have an agreement – a signed contract.

These warranties may not apply if:

- you did not make clear what you wanted done
- you insisted on having the service carried out in a particular way and you did not like the result
- you asked that the materials be used in a way in which they would not ordinarily be used and you did not like the result.

You can read more about types of contracts in the Consumer Building Guide.

Friends

Most of us have a best mate or a best friend who we have known for years and who we trust not to do the wrong thing by us. If you do have a best mate, a family member, a cousin, an uncle or anyone that you trust who is a tradesman and who is willing to do work for you, we would strongly advise against having only a verbal agreement. You simply must put any agreements or arrangements to perform any building or construction work on your property in writing, and of course it is unlawful for unlicensed tradespeople to do any work and to perform any work over the state-specified amount, including materials, without a written contract.

The written agreement does not have to be a formal contract like one that you would get from a solicitor or online from Fair Trading – it can be just a letter outlining the work, the costs, and the time frame in which they will finish the job for you. Many a well-meaning friend or relative has left owner builders in the lurch and cost thousands of dollars in delays on site, and there is always the risk of misunderstandings as to what they were going to do for you and at what price.

You do not want to test the strength of your friendship on the construction of your home. You will have a lot of other stressful things to think about without the grief and the possibility of spoiling a relationship.

Also, without a contract to perform any work you will not be insured, so you can find yourself liable for any claims, including if they hurt or injure themselves if they cause any damage to your property, or any other property for that matter.

Home warranty insurance

Home warranty insurance must be provided by any licensed contractor, builder, tradesperson or project manager who contracts directly with you if the contract is worth over $12 000. The certificate of insurance should be provided to you before the contractor takes any money and before starting any work.

As an owner builder, you need to be aware that you must provide home warranty insurance to the new owners if you intend to sell the home within a certain time of the completion date or the date on the occupancy certificate. The states differ as to the number of years that must be covered, so check your owner builder manual. A list of approved insurance providers can be found on the Fair Trading website for each state (see end of chapter).

Disputes

When you engage a builder, electrician, plasterer or plumber, you have a contract that specifies what work they will do, what materials they will use, and how long they will take to do it.

The dictionary meaning of the word 'dispute' is 'to engage in argument or debate; to argue against or call in question'. To call in question is the most appropriate meaning when we talk about disputes on building projects. We all have differing opinions on what we perceive is good quality and what we feel is defective, and in building and construction it is no different. So what is a reasonable standard of construction and workmanship?

Building disputes can arise through the owner builder not understanding the construction and building process. With this in mind and to help the owner builder determine what constitutes defective work, a collaborative paper entitled *Guide to Standards and Tolerances 2007* has been written by the Victorian Building Commission, the Office of Fair Trading NSW, the Tasmanian Government and the ACT Government. The *Guide* was prepared with the assistance of a panel of representatives from the building industry, professional associations and consumer groups with an interest in building standards and resolving building disputes. This document may be used as a guide to determine whether or not an item is defective. It is available from several websites, including that of the Office of Fair Trading NSW or from consumer regulatory agencies.

The *Guide* is intended to be used by builders and building owners to determine whether or not an item is defective only where this cannot be done by reference to the documents (your contract), the relevant Australian Standards, the Building Code of Australia, or the relevant regulations. Where there is any contradiction or difference between the *Guide* and an Act, a regulation, the Building Code of Australia or a building contract, all of these take precedence over the *Guide*. The *Guide* does not replace the requirements of these documents. It is, however, a detailed report on how a building is constructed and the problems that can arise through bad workmanship or use of unfit materials. It is recommended the owner builder print this guide out and study it for future reference.

Each state and territory provides information for the owner builder on how to deal with building complaints. The Australian consumer laws for all states is the same from 1 January 2011 and includes:

- national consumer protection and fair trading laws
- enhanced enforcement powers and redress mechanisms
- a national unfair contract terms law
- a new national product safety regime
- a new national consumer guarantees law.[1]

For further information, visit <www.consumerlaw.gov.au>.

You may have concerns with the work undertaken during construction or even some time after it has been finished. Things don't always go to plan, especially with something as complicated as building construction, so it is important to keep records and to document what has been agreed to in a contract. It is also imperative to speak to your builder as soon as you discover or suspect that a mistake has been made.

When choosing your builder/contractor it should be someone you feel will work with you, someone who will keep you updated with progress and delays, and someone you feel is approachable. Don't be afraid to ask questions on how he/she handles problems or mistakes. Make sure the builder understands that you will expect open and frank consultation about any incidents that need to be dealt with down the line. Consultation through regular meetings will go a long way in keeping the lines of communication open and aid in the resolution of issues before they become problems.

Many home owners don't understand the building process, and what looks like a blunder may not be so. Although you may see something you think is not quite right, it may well be that it is the correct installation or that there was a problem that had to be managed with something temporary until a more permanent fix could be made. Don't fall into the trap of accusing or presuming that the builder or contractor has made a mistake. It isn't helpful to assume that a mistake has been made and it will more than likely create ill feelings, which can turn into uncomfortable, unnecessary and avoidable disputes over a misinterpretation.

Or of course it could be a mistake, and depending on how bad it is it could turn into a real problem. Building your home represents a huge investment, not only in money but emotionally as well. It is far simpler to discuss problems with the builder/contractor in the hope of coming to an agreed solution. It is also a good idea to have copies of your building agreement on hand so you can refer to it, as well as quotes, invoices, photos or anything else that will help you state your case. With any luck you can both agree to an amicable resolution. Following the successful outcome of your discussion, you will need to document what you have both agreed to and the date by which you expect the problem to be rectified. Keep a copy of this agreement for your records and post a copy of the letter to the builder/contractor. If you send it by registered post you have a receipt in the event they say they did not receive it.

Mediation

If you have exhausted your attempts to discuss the matter with your builder or contractor and they do not respond, you should then contact the consumer regulatory agency in your state or territory. They will advise you on your rights and responsibilities. Although each state has a slightly different process, all the agencies have the same duties and are committed to resolving problems that deal with disputes about unfinished and/or faulty home building work and/or damage caused to other structures as a result of home building work being done.

It is a good idea at this stage to have read the *Guide* to see if the workmanship or materials used were unacceptable. If you are certain you are in the right, then start the processes and procedures. The consumer regulatory agency will take an objective look at your matter and will mediate between the two parties and attempt to negotiate towards a successful outcome.

When you contact your state agency, you will generally be asked to put your complaint in writing. You will be required to prepare a written report describing the specifics of your complaint, with copies of photos and your contract agreement as well as anything else pertinent to your claim. You will also be asked what your ideal outcome will be, how you would like the matter settled and the restitution you will be satisfied with. Don't be afraid to ask for a complete replacement if you feel repair isn't the right solution for you.

Once you have filled out the written report with all the associated details relevant to your claim, you will need to collect all your supporting documents, including any independent reports prepared along with the papers you used in the first meeting when you first discussed the building work, the building agreement, quotes, invoices and photographs of the faults/defect. You need to send all this to Fair Trading along with the application form for your claim (you can download them from their website). They will send you a letter telling you they have received the documents and give you a case number.

In most instances, a customer service officer will contact the contractor, usually by telephone, to discuss the issues in dispute and try to get you and the contractor to reach a mutual agreement. If this doesn't work you will be given a date to meet at the chambers. The day of your appointment the two parties and the mediator will discuss the complaint and listen to both sides. The mediator will ask you both to reach some sort of compromise. It is important to keep in mind that customer service officers cannot order or direct either party to resolve the complaint.

In the majority of cases the state agency is successful in helping the parties reach a resolution. Should a complaint not be resolved, the next course of action for a home owner will depend on the issues in dispute.

If the consumer regulatory authority detects that the residential building law has been broken, it will evaluate the breach and determine whether disciplinary action is required. Keep in mind that this action has no bearing on the resolution of a dispute.

Building inspector

If the complaint relates to defective or incomplete building work, it may be referred to a building inspector. The agency will have the building inspector attend your site to inspect the problems and return with a report and a recommendation.

The building inspector will arrange for both the home owner and the contractor to meet them on site. If the contractor fails to attend a site inspection, the building inspector may determine the matter in the contractor's absence.

The building inspector's role is to help resolve disputes. He or she will inspect the defects or incomplete work reported in the complaint and will then discuss their findings with the home owner and contractor with a view to resolving the issues in dispute.

The inspectors look at only those items that are the subject of the complaint. A home owner who needs a technically qualified person to do a general inspection should arrange this through an appropriately qualified private consultant.[2]

What if I don't want the contractor back?

In most instances, the regulatory agency will attempt to resolve the complaint by having the contractor return, provided they are appropriately licensed to do the work. Generally, if a contractor is willing and appropriately qualified to carry out the work, the agency will encourage resolution by way of the contractor fixing the work.[3]

If you can persuade the mediator or magistrate that you have lost all faith in the builder/contractor you may be able to contract another builder/contractor of your choice, although this outcome is unlikely.

Rectification orders

Where the inspector determines that there are defects, incomplete work or damage as a result of the contractor's work, a rectification order may be issued. The order will describe what items require rectification and a date by which the work is to be completed under the order. It will also state any conditions the home owner must comply with, such as permitting access to undertake the work. A copy of the order is given to the home owner and the contractor.

The home owner may lodge a building claim with the consumer tribunal if dissatisfied with the outcome of the dispute resolution process. If a rectification order is issued and the homeowner lodges a building claim before the date of expiry of the order, the order automatically terminates. The contractor may lodge a building claim with the tribunal if they are seeking payment of unpaid moneys from the homeowner. The contractor cannot appeal the rectification order.

If the period for rectification of work under a rectification order has expired and the contractor has failed to carry out work as provided in the order, the home owner can lodge a building claim with the tribunal. The tribunal may then make a binding determination. If the contractor fails to comply with an order of the tribunal or fails to comply with a rectification order, disciplinary action may be taken.[4]

Consumer, Trader and Tenancy Tribunal

Some types of building complaints will be referred by Fair Trading to the Consumer, Trader and Tenancy Tribunal without Fair Trading intervention. These include any complaints relating to:

- appeals against a decision of an insurer under a contract of insurance required to be entered into under the *Home Building Act 1989*
- debt recovery by a contractor
- disputes between head contractors and subcontractors for defective work
- claims arising from work done by an owner builder
- cross claims referred by the Tribunal
- claims where the time for lodging a claim is due to expire within 3 months
- matters involving unlicensed contractors.

For all other residential building complaints, the Tribunal will require the parties to participate in the dispute resolution before accepting a claim. The Tribunal member hearing the claim will be provided with the report of the building inspector where a site visit has occurred.

If the complaint relates to a contractual issue such as deposits, the home owner may be able to lodge a building claim with the Consumer Trader and Tenancy Tribunal seeking an order for the payment of moneys or another course of action to resolve the complaint.

Contractors may also lodge building claims with the tribunal for the payment of money owed by home owners, such as progress payments, additional amounts for variations or other claims for payment provided in the building contract.

Your role in the process is to supply documents and records to aid in the assessment made by the building inspector. Keep an open mind as the outcome may not be exactly what you expected.

You need to document everything and keep all your receipts and paperwork together. Document everything in writing. You may rely on it later. Keep it all together, preferably in a storage container with a lid to keep everything safe from spills etc.

Even though you may be in the right you still have to follow the procedures and processes. It is time consuming and can be expensive, as you cannot recoup all the money you have outlaid.

The Consumer, Trader and Tenancy Tribunal may make orders, including:

- the payment of money
- relief from paying money
- the delivery, return or replacement of goods
- reversing an insurer's decision on an insurance claim
- the payment of compensation for loss because of a breach of a statutory warranty, for example work not done in a proper and workmanlike manner.

The Tribunal cannot hear a building claim over $500 000. There are also time limits on making claims:

- 3 years from the date of supply on claims about building goods or services supplied
- 3 years from the date for supply on claims about building goods or services not supplied
- 10 years from the completion date of the relevant work on claims about home warranty insurance
- 7 years from the completion date of the relevant work (or if the work is not completed, from the date for completion in the contract, or if there is no such date, the date of the contract) on claims for breaches of a statutory warranty
- 3 years from the date of the contract for any other building claim.

The fact that a time limit for making a building claim to the Tribunal has expired may not prevent a building claim being made to a court.

Each side presents their evidence. If a building inspector made a report, the Tribunal may take the report into consideration. The Tribunal may also appoint an expert to advise the Tribunal. Where such an expert is appointed, no party may call another expert to give evidence unless the Tribunal agrees. Subject to an order of the Tribunal, the cost of the expert is to be shared by the parties.

Usually the Tribunal makes a decision after everyone has finished giving their evidence. Sometimes the Tribunal might want more time to think about your case, or it might direct either or both parties to provide additional documentation or clarify an issue in some other way. If this happens, the Tribunal will give its decision later. A notice of the order is sent out after the Tribunal makes its decision.[5]

How do you apply?

A form to notify the Tribunal of a dispute or a building claim can be obtained from the Registry of the Tribunal, from <www.cttt.nsw.gov.au> or a fair trading centre in your state.

You need to attach certain documents to your application, such as:

- the home building contract
- any independent building reports
- photographs showing the details of alleged defective work.

There is a fee to lodge an application with the Tribunal. Read the guide notes carefully, complete the form and return it to the Tribunal.[6]

Home warranty insurance

Home warranty insurance is mandatory for residential building work. There is some variation between states in cutoff amounts, but in New South Wales this applies for contracts entered into from 1 May 1997 with a value over $5000, and for residential building work contracted from 2 April 2002 for work with a value over $12 000. Insurance issued to 1 July 2002 covers losses from defective work for up to 7 years. Insurance issued from 1 July 2002 provides cover for a period of 2 years for non-structural defects and 6 years for structural defects. From 1 July 2002 insurance protects the home owner from loss relating to defective and incomplete work in the event that the home owner is unable to obtain compensation directly from the contractor because the contractor is dead, cannot be located or becomes insolvent. Cover for defective work under all home warranty insurance policies commences from the date the work is completed.

Home warranty insurance policies issued from 19 May 2009 onwards also enable home owners to be able to make a claim under the policy where the licence of a contractor they are using is suspended because the contractor failed to comply with a money (compensation) order in favour of the home owner made by a Court or the Consumer, Trader and Tenancy Tribunal.

A difficult resolution

I engaged a very well known furniture manufacturer to make me a leather lounge suite. When it arrived 5 months later (it was being made in Italy) the poor workmanship was astounding:

- the cushions did not match each other
- the head rests didn't match up
- the love seat with the recliner function was at least 2 cm lower than its neighbour
- the arms did not have any padding in them
- the slightest weight left big indentations.

That night I took hundreds of photos to send to the company, and the next day I called the sales person who sold me the suite. She was less than helpful.

I sent the photos and was called by their repairer who came out and looked at it, but I was at work when he came. I had to wait a couple of weeks for his so-called independent report, which was nothing of the sort. I decided not to let this company take my $12 000.00 lounge suite away to stitch up the sagging leather and put some padding in the arms, which is what the independent repairer wanted to do, because I was not going to be satisfied with them fixing it here in Australia. They would have had to take the whole thing apart.

I went to Fair Trading and got an appointment for mediation. The mediator tried very hard to get me to concede that they could fix it, but I stuck to my guns that it was the worst example of craftsmanship I had ever seen. I was very lucky to find a real independent upholsterer who had been in the industry for 35 years and taught at TAFE. He came out and gave me a real appraisal of the suite. He took photos and I gave him a hundred of my own so he could complete his report. He only charged me $120.00 – money worth spending. I sent a copy of the report to the company and a copy to Fair Trading by registered mail. I got another date some months later.

I arrived early for the hearing armed with my report and pictures and everything ready to do battle, but the company did not show. After half an hour the judge ruled in my favour and wrote orders for them to refund my money. Once the company received the order to refund my money they asked for another hearing date, claiming they had not received the original hearing date.

I was really angry at all the time it took to get to this point because I had complied with everything and sent everything by the book and taken time off work, spent money on all the postage and court hearing costs and so on. The Office of Fair Trading was very helpful and told me that if I wrote to Fair Trading with reasons why the case should not be reheard there was a possibility that they would stop the stay of order and re-issue the order. So after jumping up and down and through all the hoops I had to write another letter stating why the company should not be given a second chance. I got the ruling and Fair Trading reinstated the order.

Did I get my money back? No I did not! The company did not pay the money as ordered within the two weeks they were given. Fair Trading said that I would have to lodge the order to the Sheriff's office for them to go out to the company's place of business and demand the money or take objects for the value of the order, which of course I didn't want. I just wanted my money back so I could get a new lounge. I also found out I had to pay for the Sheriff so again I was out of pocket. If the Sheriff went there and the right person was not there to hand the order to it would mean I had to pay another fee each time they went out.

The Sheriff's offices were about 5–6 weeks behind in serving their demands for payments. I waited and waited and finally I called the court to find out what was happening. They did not know and gave me the number of the Sheriff's office. But they gave me the wrong' number. I had to find the one closest to the business. After I pleaded with the gentleman who had told me that it would take weeks to serve the order he took pity on me and they went out that week – only the right person was not there to serve the order to. I thought I would have to pay another $135.00, but to my surprise I got a call from the furniture place saying they would refund the money and arrange to pick up the lounge. They also asked me to call off the Sheriff.

Home owners must notify their insurer of:

- defects in work within 6 months of becoming aware of a defect
- incomplete work within 12 months from the date of the contract or the commencement date provided in the contract or the date work ceased, whichever is the later.

Notification within these periods will mean that the insurer cannot reduce its liability under the policy, or reduce any amount otherwise payable in respect of a claim, merely because of a delay in the insurer being notified of a loss. Nevertheless, a loss resulting from defective work can still be notified to an insurer at any time within the period of cover or within 6 months of a loss becoming apparent where that occurs in the last 6 months of the period of cover.

A home owner must take action (for example initiate dispute resolution by lodging a complaint with the consumer regulatory authority) to try to have the builder finish any incomplete work and rectify any defective building work. Where a home owner does not take action to enforce a statutory warranty an insurer may reduce its liability (or the amount paid under a claim) to the extent that the insurer's interests have been prejudiced as a result of the home owner not trying to have the builder complete or repair the work.[7]

References and useful websites

www.fairtrading.nsw.gov.au/pdfs/About_us/Publications/ft246.pdf (Consumer Building Guide) www.fairtrading.nsw.gov.au/Home_building_contracts.html

Consumer regulatory agencies

ACT

Fair Trading: <www.ors.act.gov.au/FairTrading/index.html>
Owner builder: <www.canberraconnect.act.gov.au/Services/c/cl5-constr>

New South Wales

www.fairtrading.nsw.gov.au

Northern Territory

Consumer Affairs: <www.nt.gov.au>
Owner builder: <www.nt.gov.au/lands/building>

Queensland

Office of Fair Trading: <www.fairtrading.qld.gov.au>
Owner builder: <www.bsa.qld.gov.au>

South Australia

Owner builder and Office Consumer and Business: <www.ocba.sa.gov.au>

Tasmania

Consumer Affairs and Fair Trading: <www.consumer.tas.gov.au>
Owner builder: <www.wst.tas.gov.au/building>

Victoria

Consumer Affairs: <www.consumer.vic.gov.au>
Owner builder: <www.buildingcommission.com.au>

Western Australia

Consumer Protection: <www.commerce.wa.gov.au>
Owner builder: <www.builders.wa.gov.au>

Useful contacts

National

Building Standards: Australian Building Codes Board
1 300 857 522
www.abcb.gov.au
To check if a builder or trade contractor is registered
1300 360 320
www.buildingcommission.com.au

New South Wales

Building Advisory Service NSW: <www.buildingadvisory.com.au>

Northern Territory

Northern Territory Building Advisory Services Branch (BASB): <www.nt.gov.au/lans/building/index.shtml>

Queensland

Home Builders Advisory Service: <www.hbas.net.au>

South Australia

South Australia Building Advisory Notices: <www.ocba.sa.gov.au>

Tasmania

Tasmania building advisory services: <www.wst.tas.gov>

For information about the Housing Indemnity Act 1992 contact the Office of Consumer Affairs and Fair Trading on 1300 65 44 99 or <www.justice.tas.gov.au/newca/index.htm>

Victoria

Building Advice and Conciliation Victoria: <www.buildingcommission.com.au/www/html/2540-complaints-disputes--appeals.asp>

Western Australia

A guide for owners and builders is available from <www.buildingdisputes.wa.gov.au>.

15

Exempt and complying development

When we first started out in contract housing we knew absolutely nothing about building or development control plans (DCP) and how it all worked. We soon learnt how to read a DCP and that each council has its own controls for building. Inconsistency abounded, even within the one council area. Local councils across New South Wales had up to 3100 land use zones, with up to 11 different zones in residential areas alone. As well, there were about 1700 different development definitions. There were 5500 different local planning instruments across the state's 152 councils – an average of 36 per council.[1] You just got used to reading the controls and adapting the home design accordingly, although many controls did not make much sense.

Exempt and complying development code

This complexity hindered investment in New South Wales and also made it difficult for residents to understand what sort of development could happen in their street or community. With this in mind there have been sweeping reforms to unite council development control plans within New South Wales and nationally. In this chapter we will discuss only New South Wales, and that in the most general terms. Other states and territories have either introduced or plan to introduce building regulation codes similar to the New South Wales code, but each state has its own zones and/or overlays and it would be impossible cover every one adequately. You will need to seek information from your relevant council.

The new reforms create just 25 land use zones, which councils can tailor to local needs, and the number of definitions of development is reduced to 244. This means that for the first time a definition of what constitutes development is the same in all council areas.[2]

As part of the reforms, councils will be asked to prepare a single local environment plan (LEP) covering their entire area within the five years 2010–2015. This is a planning document approved by the Minister for Planning that sets out how people can use their land and what they can build.

The State Environmental Planning Policy (SEPP) and the Exempt and Complying Development Code was gazetted on 12 December 2008, and commenced operation on 27 February 2009. The Code outlines how new residential developments on lots 450 m^2 and larger, including detached single- and two-storey dwellings, home alterations, renovations and additions, as well as demolitions, can be approved as a complying development in as little as 10 days. It also outlines how 40 types of minor developments around the home can proceed as exempt development, that is without requiring planning approval.

Development approval

In NSW the principal purpose of the SEPP was to create a more efficient code than the previous inconsistent development assessment processes (DA). In Victoria and a number of other states the planning permit (PP) and building permit (BP) are separate approvals. If after reviewing the general housing code you do not believe your development fits the required complying development controls in the SEPP, then you will need to lodge a DA, which will be a merit-based assessment. If you are doing a knockdown rebuild and you need a stormwater detention pit, you require a DA.

Complying development certificate

From 27 February 2009, those applying for development approval whose plans meet the complying development criteria can expect to have their plans approved and have a complying development certificate issued by the council or an accredited private certifier in approximately 10 days.

A complying development certificate eliminates the need to obtain a development application and construction certificate for building work. A complying development certificate combines both development and then a construction approval in the one certificate, by using key elements that address both design criteria and the technical building standards. Lengthy merit-based development assessment is no longer required and the approval process is greatly simplified.

The SEPP also outlines some areas of New South Wales state planning that are excluded from the code as well as areas where complying development is not permissible. These areas include flood or bushfire prone land development, heritage items, draft heritage items, heritage conservation areas, and draft heritage conservation areas. For a full list of the general and land-based requirements refer to the SEPP. Local councils can make an application to the Planning Minister for a local exclusion or local variation to the SEPP. So, if you are looking to submit an application for development you should check not only that you are complying with the criteria of the SEPP but also consider any local amendments that may apply to your area.[3]

Section 149 certificates

It is advisable to purchase an up-to-date full section 149 (NSW) (2) and (5) certificates. These may be obtained from the local council governing the area in which the land is located and contain a statement as to the particular zoning of a property. They also contain a table that sets out uses of the land that are permitted without development consent, uses of the land that are permitted with development consent, and uses that are prohibited. In other states you will need to get a planning certificate.

Section 149 planning certificates (NSW) also provide details of all the environmental planning instruments that apply to your land. It is important to review all applicable planning instruments as some planning instruments may modify the operation of other instruments.[4]

Criteria for complying development

The development code relies on minimum lot sizes and zones (residential, commercial, business etc.) defined by the Standard Instrument – Principal Local Environmental Plan. Land or lot requirements for complying development are set out in Table 15.1.

The general housing code for complying development allows for the erection of a new single- or two- storey home on lots more than 450 m^2 with a 12 m frontage.

The development should:

- be assessed against clearly articulated quantitative criteria or development standards
- require no judgement as to whether the criteria are met
- due to their small scale and low impact, not require public notification.

For exempt development, the proposal should:

- have low impact beyond the site
- be simple for applicants to identify
- have quantifiable parameters that need to be met
- not affect the achievement of any policy objective.

Table 15.1 Land or lot requirements for complying development[5]

Specified development	Minimum lot size (m^2)	Zone**
Internal alterations (not including the erection or conversion of a basement) – housing internal alterations code	Any lot size	Any zone
New dwelling houses – general housing zone	450 or greater	R1, R2, R3 or R4
Alterations and additions – general housing code	450 or greater	R1, R2, R3, R4, R5, RU1, RU2, RU3, RU4 or RU5
Ancillary development* – general housing code	450 or greater	R1, R2, R3, R4, R5, RU1, RU2, RU3, RU4 or RU5
Demolition of an existing single-storey or two-storey dwelling house or ancillary development – general housing code	450 or greater	R1, R2, R3, R4, R5, RU1, RU2, RU3, RU4 or RU5

*Ancillary development is defined as any of the following that are not already exempt development under the code:
(a) access ramp
(b) awning, blind or canopy
(c) balcony, deck, patio, pergola, terrace or verandah that is attached to a dwelling
(d) carport that is attached to a dwelling
(e) driveway, pathway or paving
(f) fence or screen
(g) garage that is attached to a dwelling house
(h) outbuilding
(i) rainwater tank that is attached to a dwelling house
(j) retaining wall
(k) swimming pool or spa pool and child-resistant barrier.
** Equivalent zones. The Codes SEPP relies on the land use zones (i.e. residential, commercial, business etc.) established under Standard instrument – Principal Local Environmental Plan.
Source: A Guide to Complying Development: Housing. State of NSW through the NSW Department of Planning, May 2010, Table 3.

The key controls to be assessed include:

- zoning
- lot size
- site coverage
- building height allowed per zone
- setback requirements
- secondary setbacks for corner blocks
- side setbacks
- rear setbacks
- landscaping
- outdoor area for private open space. These are listed in the New South Wales housing code.

Since this criteria is for NSW, owner builders in other states should consult with their local council who can provide development control plans explaining the requirements for their block size and the correct setbacks, floor space ration and the like.

Table 15.2 Key controls from the SEPP for different lot types[6]

Lot type	Lot size range	Lot width minimum	Site coverage maximum	Floor area maximum	Building height maximum	Front setback minimum: average of the nearest two houses or if no house	Secondary street setback minimum	Side setback minimum	Rear setback minimum (m)	Landscaped area minimum	Principal private open space minimum
A	450 m^2 or at least 600 m^2	12 m	50%	330 m^2	8.5 m	4.5 m	2.0 m	0.9 m plus calculation	3.0 m plus calculation to 8.0 m	20%	24 m^2
B	600 m^2 or at least 900 m^2	12 m	50%	380 m^2	8.5 m	4.5 m	3.0 m	0.9 m plus calculation	3.0 m plus calculation to 8.0 m	25%	24 m^2
C	900 m^2 or at least 1500 m^2	15 m	40%	430 m^2	8.5 m	6.5 m	3.0 m	1.5 m plus calculation	5.0 m plus calculation to 12.0 m	35%	24 m^2
D	Over 1500 m^2	18 m	30%	430 m^2	8.5 m	10.0 m	5.0 m	2.5 m plus calculation	10.0 m plus calculation to 15.0 m	45%	24 m^2

Source: A Guide to Complying Development: Housing. State of NSW through the NSW Department of Planning, May 2010, Table 4.2.

Table 15.3 Site requirements: maximum site coverage by lot type[7]

Lot type A	Lot type B	Lot type C	Lot type D
50%	50%	40%	30%

- The total area of the lot to be covered by a dwelling house and all ancillary development (e.g. carport, garage, shed) varies from 30% maximum to 50% maximum depending on Lot Type.
- The calculation of site coverage does not include access ramps, awnings, eaves, unenclosed balconies, decks, pergolas, terraces, verandas, driveways, farm buildings, fences and screens, rainwater tanks attached to the house, swimming pools, spas or development where the Exempt Development Code applies.

Source: A Guide to Complying Development: Housing. State of NSW through the NSW Department of Planning, May 2010.

There is a range of maximum floor areas for ancillary structures depending on their use and the lot's zoning: see the SEPP for details.The code appears to be rather generous in the size home you are permitted to construct, as the DA for many councils allows a floor space ratio of only 0:045 (45%) and this equates to a pretty small home by today's standards. For instance our block of land is 690 m^2, and with an FSR of 0:045 we can build only a 35 sq home including the garage, but if we used the Code we could build on 50% of our block, to around 38 sq. Many councils include a swimming pool in the FSR calculation, so if we wanted a swimming pool we would have to reduce the size of our home to accommodate the pool.

Complying development essentially provides for a tick-box assessment of a proposal against a set of criteria. You can download the complying development checklist from <www.planning.nsw.gov.au>. This checklist has been developed to assist you when submitting your application and contains information that is in addition to the criteria specified in the codes. Some councils and certifiers may require additional information.

For a more detailed explanation of how the system works:

1. Contact your council: many councils have established special units to handle complying development and they will be able to help you understand how to submit applications that comply with relevant complying codes.
2. Contact a private certifier.

Table 15.4 Site requirements: maximum floor area by lot type[8]

Lot type A	Lot type B	Lot type C	Lot type D
330 m^2	380 m^2	430 m^2	430 m^2

- The floor area includes any carport, garage, balcony, patio, pergola, terrace or veranda attached to the house and with an enclosing wall of at least 1.4 m above floor level
- The maximum floor area of a house must not exceed the areas in the Floor Area table
- The calculation of the dwelling house floor area is the total of both the ground and upper floor (if there is one) not including awnings, eaves, voids, stairway or lift shaft

Source: A Guide to Complying Development: Housing. State of NSW through the NSW Department of Planning, May 2010.

Private certifiers

Private certifiers are able approve your application for a complying development certificate provided the proposal meets the criteria in the relevant code. Certifiers can assist you in determining whether or not your proposal can be approved as complying development. Chapter 17 discusses private certifiers/building surveyors etc.

Certifiers are registered through the Building Professionals Board (BPB). If you want more information on how to contact certifiers you can visit the website of the BPB at <www.bpb.nsw.gov.au>.

You do not need a construction certificate where a complying development certificate has been issued for the plans and specifications for the work. However, you will need to obtain your construction certificate or complying development certificate before commencing building or subdivision work or you will not be able to obtain an occupation certificate.

You must also appoint either a private certifier or take the forms to council for assessment. Your private certifier will inspect the building work during construction; once they are satisfied it meets the code they will issue an occupancy certificate.

Exempt development code

The new planning system has introduced an exempt development code for minor developments. This comprises 40 different types of development deemed to be low impact, and these therefore can be undertaken without the need for planning or construction approval. However, other legislative requirements for approvals, licences, permits and authorities still apply.

These minor developments include small-scale structures associated with a dwelling such as sheds, pergolas, fences, rainwater tanks, fixed barbecues , access ramps, air conditioning units, animal shelters, awnings, decks, patios, pergolas, balconies and verandahs, just to name a few. You will find a comprehensive list of exempt developments on the Department of Planning website.

Generally, the exempt development code applies to all land in New South Wales, regardless of zoning or lot size. It is based on a review of standards previously used by councils across New South Wales and is designed to provide a reasonable and consistent set of development standards.

Future changes

There is talk about more reforms for:

- simplifying some existing development standards and allowing minor external alterations, such as enlarging windows and doors

- bushfire affected land or flood affected land.

Check with your local council to see if you are able to take advantage of these changes.

The Department has extended the complying development code for lots of the following sizes:

- 200–249 m^2 maximum 65% footprint
- 250–299 m^2 maximum 60% footprint
- 300–449 m^2 maximum 55% footprint.

Owners of these smaller and often narrower blocks will find this new legislation helpful in dealing with design issues, as you must be mindful of many problems with solar access and the impact on your neighbour's homes, and more so if you want to build two-storey home because of overshadowing and privacy issues.

You can find more information on this new legislation at <www.planning.nsw.gov.au> or check with your local council.

A choice of processes

Summing up, there are three approval processes the owner builder can utilise:

1. Exempt development, mainly for low-impact renovations and minor things like a shed or swimming pool.
2. Complying development, which covers low-impact and routine construction.
3. Merit-based or standard DA application.

The primary purpose of the new code is to create a faster, more streamlined and cost-effective approval process while reducing red tape for those complying developments that fall within its framework. This saves time and frees up council planners for bigger development applications like infrastructure. There is also an added bonus of savings of around $6500 consisting of reductions on mortgage payments and holding costs, not to mention council fees. This means more money to spend on those items you may have wanted but thought you could not afford.

References and useful websites

Australian Building Codes Board: <www.abcb.gov.au>

New South Wales

Building Professionals Board: <www.bpb.nsw.gov.au>
Department of Local Government: <www.dlg.nsw.gov.au>
Department of Planning: <http://housingcode.planning.nsw.gov.au>
email: codes@planning.nsw.gov.au

Environmental Planning and Assessment Act 1979
www.legislation.nsw.gov.au

Northern Territory
The Northern Territory Planning Scheme
www.dpi.nt.gov.au

Queensland
Integrated Planning Act 1997
www.dip.qld.gov.au

South Australia
Development Act 1993
www.planning.org.au/sa

Tasmania
Land Use Planning and Approvals Act 1993
www.planning.tas.gov.au/the_commission/about_the_commission

Victoria
Planning and Environment Act 1987
www.dcpd.vic.gov.au/planning

Western Australia
Town Planning and Development Act 1928
www.planning.wa.gov.au

16
Theft and vandalism

Cost of theft

Theft and vandalism affect nearly every construction site across the world and costs billions of dollars annually. The high cost of building materials induces some people, including some contractors, to steal materials from construction sites in order to reduce their own building costs. Theft is committed inherently for the money; the decision to steal is influenced by the perception of the ease with which the crime can be committed. Thieves often know their victims, who can include acquaintances, neighbours, relatives, friends, or friends of friends and relatives. They do not give much weight to the consequences of their actions or believe they will be caught. In fact they are right. Construction site thieves are very rarely caught.

Construction sites are targeted for tools, equipment, machinery and building supplies, and theft adds approximately 1% to the cost of building a home. This additional cost comprises not only tools and building material, but also the cost of downtime, loss of production and increased costs in insurance premiums and increased cost for hired equipment.

The high cost of theft and vandalism incidents on construction sites can obviously have a very negative impact on the success of your project. Site security is often overlooked or not taken into consideration by owner builders until after the theft or vandalism has occurred. You should include in your construction budget an allowance for the security of your site.

Security measures

There are many different types of security measures for preventing theft at construction sites. These include locks, gates, alarms, security lighting, cameras, and engaging your neighbours and residents for their support in keeping an eye out to report anything suspicious. We would stress to them not to interfere with the intruders – ask them to call the police instead.

Surveillance cameras

An effective technique is installing hidden surveillance cameras. The cameras available today are not too expensive and are virtually undetectable. There are also security systems you can hire. One system we found on the internet is easy to set up and designed on a trailer that can be moved easily onto your site. It has its own solar-powered generator and infrared motion-sensor camera. This can be brought onto your site and gives you 24-hour remote access to live and recorded video. It costs around $2500 per month, but you can use it to monitor the whole construction site. An added bonus is this camera includes work place health and safety monitoring.

Fences

It is a good idea to have your construction site fences installed before any equipment arrives on site. This keeps nosey passers-by out, and is the first line of defence against thieves.

The best form of fencing is cyclone or chain-link fencing; you can hire this. It must be secured in precast concrete blocks or anchored base plates on the ground, and it will be installed by the hire company. Chain-link and cyclone fences allows for surveillance by you, police and neighbours. You could also look into barbed wire at the top of the fence. This makes climbing much more difficult and uncomfortable. It also acts as a psychological barrier, and the higher the fence the better. It is even more effective if you attach the barbed wire to a bracket tilted at a 45° angle.

Your cyclone fencing is the perfect place to put up your licence and contact details along with phone numbers for the police and fire brigade. Post signs along the fence that prohibit trespassers and warn of danger. You could also advertise surveillance cameras are in use on your site even if you don't have them. It will make any would-be vandal or thieves think twice.

Lights

Fencing should not be your only security or safety measure. Bright white lighting can be an effective deterrent to theft and vandalism on site. It is particularly effective in deterring the impulse or casual offender if they think the site is being monitored, so keep the construction site well lit at night or have motion-sensor spot lights. We recommend you visit your site at night so you can see if there are any possible hiding spots for thieves so you can rectify positioning of lights.

Visibility

Make sure the front of your construction site has a clear zone of at least 4 m. Remove any vegetation and have construction rubbish or skip bins stored away from the fence. This aids passive surveillance and means would-be thieves cannot use trees or construction material to help them to climb over the fence.

Loss prevention measures

You should also discuss loss prevention measures with your contractors. It requires effort on everyone's part, but if you do not make it a priority neither will they. Do not make it easy for a would-be thief to enter your site pretending to be a sub-contactor, as thieves use this tactic to get to know the area so they can come back when the workers are gone. Most theft and vandalism takes place after the workers have left for the day and on weekends when there is no one around.

Develop a system to verify the identity of people who are genuinely working by asking open-ended questions like 'Who do you work for here?', 'What do you do for that contractor?', 'When did you come on this site?' Ask specific questions that only an authorised person would know about the job. If the visitor hesitates or becomes defensive the chances are they do not belong there.

Have a pre-construction meeting with your contractors and write a security provision into the contractors' and subcontractors' agreement. This should identify who is responsible for job loss on your site, establish theft prevention policies, and show the steps you have taken to demonstrate to them that you are taking an active interest in reducing theft and vandalism.

Deliveries and secure areas

Be sure to also discuss who will be responsible for deliveries being made to the site if you are not available to receive them. Make sure there is an area that is secure with double cylinder dead bolts and a monitored alarm for all your tools, machinery and materials, and have each of them engraved with your licence number or logo on them. You may also want to tag it stating you have all serial numbers recorded and logged. This will make the tools harder to sell and easier to identify if recovered.

At this stage of your development it is crucial to get security right, as before the doors and windows are in place and you can lock it up your house is very vulnerable to thieves who can enter and remove your copper pipes and electrical wiring and anything else they can find. Do not have your white goods, hot water systems, ducted air conditioners, kitchen appliances, kitchen, or anything that is easily dismantled and removed from the site before the house is secure and/or you have moved in. You could consider an on site storage container with padlocks that are not easily cut with bolt cutters. Coordinate deliveries of materials and appliances so that they are delivered close to the time they will be installed or the house will be occupied.

If you are robbed

Make sure you report any thefts to the local authorities even if you are not going to put in an insurance claim. (Excess costs may mean it is not worth making an insurance claim.) You can also do the rounds of markets, car boot sales and hock shops to see if your tools turn up there. Having recorded the serial numbers should make it easy to recover them.

Nightmare before Christmas

About two days before Christmas and the long summer break for the building contractors I received a call from a client, who had been to see her almost-finished house.

My client told me she had stopped a man with a ute from taking the cyclone fences from her construction site. When she challenged him he said 'I am borrowing the fences. We builders always borrow from each other.' She told him he couldn't take the fences. Although she was very scared she took photos of his numberplate because he ignored her and took the fence to his vehicle. He was really menacing and told her to stop taking photos. She had her two young children in her car and got in and locked the doors. She watched him put the fence back and called me. She was worried that he would hurt her or the children because she had the photos. I told her to go home and not worry, because he would not do anything to her as she had evidence of what he was stealing.

When the tradespeople arrived back on the job on 19 January they found that an upstairs tap had been left on in the bathroom and the whole house had been flooded. The top and bottom floors had to be ripped out and all the internals had to be replaced. This obviously cost the builder money in an insurance claim, and lost money in production time. The client had to rent for a further 6 months while everything was rebuilt. The client was too scared to tell the builder about the man she stopped and the photos in case he came back to hurt her when the house was finished. I didn't tell my company about this incident either, as she was extremely scared of what could happen to her or her two young children. I personally felt responsible for her safety, and the company had insurance. We did not have evidence it was the fence thief, so nothing could be gained by telling management about the incident.

Another time at a new home site vandals got in by breaking a window. They found cans of paint and emptied them on the floors and splashed it on the walls. They also kicked in some walls. Fortunately in that case the damage was easy to fix.

Vandalism

Vandalism is generally caused by teenagers or aggrieved or disgruntled employees. Vandalism can create weeks or months of costly extra work on your construction site, and should not be ignored. It isn't uncommon for vandals to hit the same site several times. Installing closed-circuit security cameras, providing floodlighting and following the anti-theft suggestions to secure your site will go a long way towards saving you time and money.

17

Certification

Private certification

When the time comes to submit your development application for approval you have a choice to submit directly to your local council building surveyor or to retain a private building surveyor. Legislation has been introduced to enable competition and allow private building surveyors to issue building permits. This system is known as 'private certification' (NSW).

You may know these as Building Services Authority/Building Commission or Building Practitioners Board in other states and territories.

A private building surveyor is an accredited professional who is authorised to issue certificates for development approval. The building surveyor is able to check that development proposals comply with required technical standards and regulations of the Building Code of Australia in the same way as your development application would be assessed through your local council.

If you engage a private building surveyor you still have to comply with the council's rules and regulations. An accredited building surveyor is obligated to a strict code of conduct, and has a duty always to act in the public interest. Serious penalties can apply if they fail in these duties.

The Building Services Authority carries out audits of the building surveyor's work, investigates complaints, and can take disciplinary action against building surveyors found guilty of professional misconduct. Any person may lodge a complaint with the Building Services Authority against the actions of a building surveyor.

If you appoint a private building surveyor, under the Integrated Planning Act 1997, the engagement must be in writing and must state the certification fee.

Private building surveyors also require professional indemnity insurance as they are hypothetically responsible to make good bad or substandard work if it can be demonstrated that they have been negligent.

Responsibility of the building surveyor

A private building surveyor or local council building surveyor is responsible for:

- issuing construction certificates, certifying (among other things) compliance with the Building Code of Australia (BCA) (NSW). For the other states the construction certificate could be called a building permit or the building approval.
- issuing compliance certificates specifying that conditions of consent have been satisfied or that work complies with the plan and specification, or nominating the classification of a building under the BCA
- issuing complying development certificates, certifying that nominated development proposals comply with standards and criteria in council's local environmental plans (LEP) and development control plans
- acting as a principal certifying authority (PCA), responsible for issuing, among other things, occupation certificates specifying that buildings are safe to occupy and subdivision certificates specifying a subdivision can proceed to registration where the council's LEP permits private sector involvement.[1]

Thus a building surveyor, private or council, issues the certificates for each step of the construction to proceed to the next phase. A building inspector cannot issue the certificates.

Construction certificate

To commence works on the site you need a construction certificate. Again you have a choice to engage either the council or a private building surveyor to certify the construction of your development.

The construction certificate can be issued only after:

- council has issued development consent for the same development
- specifications and information have been provided with the construction certificate application to ensure compliance with the relevant standards including the Building Code of Australia
- where relevant, you have complied with specific conditions of the development consent.

Principal certifying authority (NSW)

Once you have received a construction certificate, you will need to engage a principal certifying authority (PCA). Again, you have the choice of either council or your private building surveyor to be your PCA. The PCA is responsible for:

- overseeing the construction works on the site
- ensuring that the relevant conditions of the development consent are being complied with
- ensuring that stages of the construction have been duly certified by the appropriately qualified professional
- issuing an occupation certificate for the building before the building can be occupied or before use of the development commences.

If council is not your principal certifying authority, you are responsible for advising the council of your nominated PCA, including their details, 2 days before you commence construction works. To ensure that your development is completed in a coordinated and timely manner, you are strongly advised to engage the person who is issuing your construction certificate to also be the PCA for the construction phase.[2]

Compliance

Inspections for compliance with approved work must be carried out by either a local council building certifier or a private building surveyor. The certifier who approves the plans for building work must also do the required inspections personally or authorise a competent person to do them.

Note that while the builder has a statutory obligation to notify the certifier at certain construction stages, home owners are ultimately responsible for ensuring that approvals are granted and the required inspections are carried out. Arrangements for certification should, therefore, be discussed and agreed with the builder and recorded in the building agreement. You are entitled to copies of certificates of inspection at each stage.

The building surveyor often will engage a building inspector to undertake some or all of the inspections required by law. Because building inspectors have particular expertise in the construction aspects of a building, they are well placed to advise the building surveyor on the level of compliance being achieved on site. However, unlike building surveyors, a registered building inspector is not able to issue building permits.

Building permits are given on condition that certain mandatory inspections must be carried out. Once the building work reaches a stage when the work can be inspected, the builder must give the building inspectr a notice, whether in writing or by other means, advising that building work has been carried out to a stage when inspection can take place.

The purpose of the inspections is to ensure that the building work is being carried out in accordance with the building permit and relevant building standards.

Building inspections for the construction of a house must be performed at the following stages:

- Footing: includes inspection of the foundation material and the reinforcing steel before concrete is placed.
- Slab: includes a check on the bearing capacity of the soil, inspection of the moisture-proof barrier, and the reinforcing steel before concrete is placed.
- Frame: covers inspection of the frame including timber sizes, fixing, tie-down and bracing before the cladding or wall linings are fixed.
- Final: includes checking on any previous outstanding items and the collection of certificates such as termite protection, wet area membrane installation, glazing, and certification of engineer-designed elements such as roof trusses.

The final inspection will also cover aspects such as:

- the control and discharge of stormwater
- the height of the floor above ground
- support of any earthworks necessary to protect the building and other property
- protection against water penetration
- fire safety issues such as smoke alarms, solid fuel appliances, fire protection near boundaries and in designated bushfire areas
- room ventilation
- toilet door swing
- vermin proofing
- sub-floor ventilation
- termite protection and re-treatment notice
- stairs, handrails and balustrades
- swimming pool fencing.[3]

18

Contract builders

You have worked very hard to be in a position to build a new home. That is why it is important to find out how it all works, how you need to be involved and how to take ownership of the job at hand, even if you decide to engage a contract builder rather than being an owner builder.

The quest starts with finding the right home design that suits you, your family, your lifestyle, your budget and your block of land. As a start you can hit the display villages and websites of prospective builders to look at project homes. You can also get inspiration from magazines. You might even put all the ideas you like together and design your own home.

The salesperson

Your first point of contact when visiting a display home is usually the salesperson. There are many details the salesperson has to be familiar with: they have to be knowledgeable in all aspects of development applications, permits and approvals.

The salesperson's job is to give you all the necessary information about the homes they have and what problems if any may crop up in conjunction with the building. A good salesperson should tell you what is realistically possible on your site and what is not. The more investigating you do before visiting display homes the more it will help you decide whether the salesperson is for you.

Pricing a project home

A display home will generally have two prices associated with it: the base price and the 'as displayed' price. In other words, the house you see on display may contain features or fittings that are not included in the house for the price you are being quoted. Check the base level of inclusions to find out what you are actually getting and compare the homes you are considering on that basis.

Features of a display home that are not included in the price may include flooring, window coverings and light fittings. Find out the costs of such features and factor them into your budget. It is extremely difficult for the salesperson to quote a price for 'this house' because of the promotions and different levels of specifications built into each home.

The majority of project builders will offer you some sort of promotion, usually a number of fabulous upgrades for a very low price, such as a free alfresco, free ducted air-conditioning, popular types of kitchen cabinetry such as Poly Kitchens and Caesar stone tops, and upgraded taps, door furniture, or front entry doors. These upgrades, usually worth about $7000–$10 000, are already costed into the base price so when you pay your nominal $3500 to get the upgrades you have in fact paid full price for them.

Remember the choice of building materials is usually limited. You may need to pay extra to get a type of brick you like. Check what options are included in the base cost. Check what sort of price range is attached to upgrading the features that are particularly important to you.

The final cost of the home is also affected by your block of land. For example, the slightest slope to the ground will require cutting or filling. This cost, called a site cost, is not included in the base price of your home. Any significant fill can create a need for additional piers to be placed into the ground – another extra cost.

If you are doing a knockdown rebuild you must clear the building envelope ready for the slab. This should be done just before you seek council approval for construction. The majority of project home builders do not want to get involved with demolition and clearing your site, and removal of trees will be an extra cost.

Some builders may not build in your area. Some are willing to travel at no extra cost but others may charge a travelling allowance – even if your site is only 15 minutes outside their standard area it can cost an extra $3000–$15 000.

Design modifications

If you decide on a project home, you need to ask if the design meets all your needs or if it needs tweaking. Ask the salesperson if they allow changes, as not every builder will allow it. For the bigger builders, volume is more important than customising your home for you. If you don't make changes and you stick to their original plans, they can churn the house out more quickly. Other builders are very flexible as long as you have the budget to pay for it.

The reason builders don't like you to make changes is because it takes longer to process (drawings take longer, as they have to be drawn from scratch instead of from their database of designs), it takes longer to price if you are changing room sizes and roof spans and the bill of quantities changes, it costs more to build, and you may be asking for features they do not have the expertise to pull off. Normally project home builders design their homes to be constructed without difficult roof lines for ease of building. They also usually have a few different façade options (at an additional price) that can be adopted to the design.

Whichever way you choose the design, you will need to take it to someone who will draw it up to scale and make sure it fits on your block with your council controls, and also those of your estate if there are any. Some project home builders will let you make an appointment to sit with their architect to design your home. You will incur a charge of about $2000, which is really a small price to pay unless you can get it free from the consultant. However, be prepared that this takes time. Depending on the limitations of your block you may have to compromise. If you are using your own design you will need to take it to an architect or a draughtsman to draw it up properly.

Whether you are using your own design or a project home design, you will need to get it priced. A lot of project home builders will build your design, but it usually costs more per square metre than their standard designs. Some builders will charge a lot more; others aren't too bad.

Choosing your builder

You need to be assured that your builder will be beside you at all times, a partner to help support you through any potential issues, from technical matters to local government requirements or construction principles and techniques. You are placing an enormous level of trust in your builder of choice to guide you through all aspects of the building process, from beginning to end. The builder has total control over your development, when it starts, when it finishes and everything in between, and you will need to be prepared for all the ups and downs.

We have heard many stories about project home builders. Some projects have had nothing but problems and others have run smoothly. Most contract home builders will have a few obligatory names and pictures of happy clients on their website. Because of privacy issues most are unable to provide you with names and numbers of clients for you to speak to about their experience with the builder, although you should also keep in mind that the builder will never tell you about the unhappy clients.

You should do as much research on your prospective builder as possible, and the best place to start is the internet many people document their experience on blog sites or product review sites. Many prospective clients ask to see a home under construction, but this is generally not useful as most people don't know what they are looking at.

To get help deciding on a builder look at <www.productreview.com.au> (type in the name of the builder) and blog sites, and listen to word of mouth, friends, relatives. Also you can call the Master Builders Association for a shortlist of builders in your area doing the type of work you require. The Housing Industry Association (HIA) will be able to give you a similar list. Look for advertisements on television, radio, in home buyer magazines or the local newspaper. Look in the Yellow Pages under Building Contractors, and under the various types of trade categories. The Office of Fair Trading also has information on how to select a builder or tradesperson.

The tender or quote

Once you have settled on the design the project home builder will do a tender or quote.

There are a few different ways that the builder can give you the cost to build your home.

Budget quote

The builder will not charge you a fee, but will give you a budget quote that will not give you a true cost to build your home, with site costs and connections to your services, although it may have a basic cost to meet your sustainability requirements. This type of quote is mainly used if you have made a lot of variations to the design and you need to cost it to make sure it is within your budget.

Comparative tender

A comparative tender can also be done free if you already have a tender from another builder. This is a really good way to get the builder that you prefer to give you a tender that isn't inflated, because they want to win the job, but you have to be careful that they haven't skipped something to make their price look better.

You will have to give the builder the other builder's complete tender with contours, soil tests, estate requirements, council requirements, and standard sustainability assessment (some councils require different size rainwater tanks, for example, and obviously this will incur a different cost). Your real sustainability assessment will not be done until you have paid a deposit and your architect drawings have been done, but before your development application goes to council or if you are applying for complying development.

The comparative tender will include the variations and inclusions you want in your home with your choice of façade – for example if you want render or bag and paint or stacker stone. Most builders will have some sort of free offer or a promotion, so if you are comparing builders you will have to work out what is standard, plus the value of the promotion or free offer, compared to what you would get with the other builder.

Some builders advertise that they give you more in their standard inclusions than other builders. These days there really isn't that much in these claims, although there are still some builders who are giving only single power points in the bedrooms. Some builders have different levels of specifications on top of the promotions they are offering, usually budget, medium and higher level. They will have associated costs for each level, so construct a spreadsheet to compare what each builder does and does not include.

It can take a lot of time working out the differences between two tenders, but it is usually worth it because that extra work helps decide who had the better finish and inclusions.

Rough site costs estimate

The builder may charge you a nominal fee, say $250, to do laser levels on your block. This will determine how much slope there is and if you need to modify your foundations such as by using deepened edge beams, or if you will need to build a split-level home.

The tender may include a provisional allowance for spoils, the excess soil that has to be taken away when they excavate for the slab. If it doesn't then you could be up for more money. Or you may have to import soil instead to get a flat bed for your slab.

The site costs estimate usually gives a rough idea of the site costs to build your home, but it is in no way what you will actually pay as there is not enough detailed information on the site to give the builder the ability to work out the real cost. One builder we worked for did not even bother to go out on the site to do the levels – instead they went to Google Earth and typed in the address and looked at the site that way. You will not get a true cost to build your home by this method.

Sometimes the builder will tell you if you need retaining walls, although these are more often constructed by the owner when the development is finished and the need depends on your foundations, on the slope of your block, and how much cut and fill is allowed by the council.

Detailed site costs estimate

The next and possibly the best way to get a quote is to pay a reasonable fee, usually $750–$1500, depending on how many reports the builder does. Ask what you get for the money. The fee will be counted as a deposit towards the total cost of the home, so if you do decide to engage the builder concerned you are not out of pocket. Ask the builder to post the contour and geotech reports to you so you can use it to inform other tenders without having to pay the fees again.

For this fee you will get a soil test, which determines your slab construction. The engineer uses the soil test to decide how much piering you need and how deep the piers need to be. If you have a P or problem soil it can cost another few thousand dollars, as they have to keep on pouring concrete into the bore holes until the piers do not collapse.

For this money you should also get:

- contour levels (this is called a contour survey)
- planning certificates (149 certificate, 88B certificate NSW), title searches, water board checks to show where your sewer is located (you do not want to build over the sewer as it would have to be encased in concrete and therefore cost extra).

For a knockdown rebuild you need a curb and gutter report, and overhead power lines must be noted in case the builder has to use a crane to deliver the frames and trusses or other materials.

If you are on a busy road do you need traffic controllers so trucks and supplies can be brought in safely? Is the site big enough to store materials or will you have double handling fees? Will you need a borehole to connect you to your water service if the water supply mains are the other side of the road?

Provisional allowances

The builder preparing a detailed costs estimate will include a provisional allowance for some items but will not prepare reports until they win the tender.

- You may have to have a landscape plan, and you will require an acoustic report if you are in flight path or near a railway line.
- If you are in a flood zone, do you need to be above the tidal wave? If you do you will need bearers and joist construction.
- If you are in a mine subsidence area you will need extra piering and possibly a thicker slab.
- You may require additional reports if you are in a bush conservation area or bushfire attack zone.

Fixed price contracts

The majority of project home builders will promote a fixed-price contract. The fixed price relates only to the building of your home. The cost of the house may not go up if or when the builder makes a mistake, but if they promise that you will never have to put your hand in your pocket for anything they will have included a buffer to cover unforeseen expenses, including additional site costs. You will be paying for the unknown whether it actually happens or not, and this can increase the cost of the home by $4000 or more.

Builders cannot afford to lose their margin – if they did they would soon go out of business. No builder can give fixed site costs because they don't know how much piering will be required until the day they do the piering for the slab. For this reason there will be a provisional allowance in the tender. Often the site costs will go over the allowance, for example if they need more concrete than the engineer estimates, or they hit rock when they are drilling. There will also be provisional allowances for excavation, cutting and filling.

Similarly, the builder cannot fix your costs for complying with sustainability requirements, as this can change according to your solar aspect, the homes next to you, your windows, and many other factors.

Start and finish dates

Project home builders promise a guaranteed start date. This is dependent on a host of things including relevant approvals and finance.

Some project home builders promise that they will complete your home in 18 or 26 or 32 or 36 weeks, but this guarantee is again subject to a host of disclaimers and clauses in the contract. Make sure you read the fine print and the builder's annexure in the contract. Try to have a flexible rental agreement if you have to rent during the construction as builders rarely finish on time. The longer it takes to build your house, the longer you may be paying rent.

The builder includes clauses in the contract for lost days due to rain, and lost days due to public holidays and the Christmas break, which is usually four to five weeks.

A project home builder may not guarantee the build time if your home has any variations at all from the standard design, such as being subject to a split-level design, home sites with more than one metre fall, if you are doing knockdown rebuild, if they have to use traffic management. Even some standard inclusions such as stone benchtops may pose scheduling problems.

The contract may give you a rental allowance if your home isn't finished when promised, but when you add up all exempt reasons why the house couldn't be finished on time the liquidated damages are negligible.

Structural guarantee

Your contract will also have a structural guarantee of anywhere between 7 years to 50 years, depending on the builder. The structural guarantee covers major structural element defects, attributable to defective design or workmanship. This includes situations where the defect results in the building or any part of the home becoming unusable, physical damage to the building, or the building posing a threat of physical damage or collapse.

Structural elements of your home these are those elements that are essential to the building's stability and structural integrity, namely, the frame, roof trusses, slab, piers and footings.

There are some conditions under which your guarantee becomes void. These include if you have carried out structural alterations or additions to the property, if you have transferred the property, or if the damage has been caused by a lack of reasonable maintenance or action to minimise damage. You should take care to maintain your home over its lifetime, including major maintenance activities such as replacing roof tiles after their useful life. This will protect the structural integrity of your home and minimise the likelihood of your guarantee becoming void.

The everyday ageing of your home is not covered by our guarantee. This includes shrinkage and minor settlement cracks and normal wear and tear. Damage caused by an act of nature exceeding that allowed for by the relevant Australian Standards or good building practice is also not covered.

Deposit

Once you have agreed on the tender you will pay a deposit. It is usually of the order of $4000 but may be as much as $10 000. This is to pay for the architectural drawings and is a sign of commitment from you that you are ready to proceed to the construction phase.

Signing off on the plans

The plans usually take 3–4 weeks to be drawn. Once your plans are ready you will be invited to take them home to study. You must also check that everything you have asked for is indeed on the plans. when you are ready take them back to your consultant she/ he will help you to make any adjustments that are required and it then goes back to the architect to be redrawn depending on how much has to be changed it usually takes a week or two. You will be given another appointment to review the new plans and if all is well you sign them off.

The contract

Each builder has a different order as to what happens next. Generally they will prepare your contract to be signed. The contract includes your tender and your plans. You can take it to a solicitor, but HIA contracts are pretty easy to read these days. If you don't understand something you only need to ask the sales consultant you have been dealing with to explain it, and if you don't like the explanation then take it to someone you trust to clarify any concerns.

You will also pay the remaining balance of your deposit. This is usually 5% of the total price of the contract less any fees you have already paid, if you haven't already organised your home loan now is the time.

The contract will mostly be to the builder's advantage, so read the fine print carefully. Most builders put in very long completion times and will always err on the builder's side as will most of the contract. Please note that if you do decide to build with a contract builder you will in most cases have to use only the trades and services provided by that builder. You can shop around to find a builder who is flexible, and if they are, you must have everything included in the contract so that there can be no misunderstandings about who agreed to what. It is also fair to say that you will be relinquishing any independent input into your development once the contract is signed. Make sure you are very thorough in reviewing the contract as it might save you a lot of anguish later. Building a home can be extremely emotive.

Sustainability assessment and approvals

Your next step is to choose your external colours. Do this while you are waiting for your sustainability report. You may also do your internal colours and your electrical layout at this time if that fits in with the builder's procedures.

The sustainability assessment usually takes a couple of weeks to process. In some cases you may need to make some modifications to satisfy the assessor. The next stage is council submission. If you are building in a new greenfield estate the builder will lodge your application to the land developer before sending it to council for approval. This is to satisfy the developer that you have incorporated design elements into your home that are necessary for the estate. Modifications to meet estate design requirements add to the cost of the house and you may be charged extra for reissuing the architectural drawings. Depending on the modifications required you may need a new sustainability certificate.

Approval time in council for a development application may vary between eight and fifteen weeks, and in some council areas may take even longer. You will also be invited to finalise your colour selections they will work with you in choosing your inclusions and internal finishes. It is important that you are familiar with your choices and colours for your new home before the meeting as this will assist in the outcome it takes at least two hours to complete and in most cases you need to do it during business hours so you may have to take time off work.

If you are doing knockdown rebuild you will need to have your demolisher on board to submit your demolition request to council for approval. However, we would not knock down until you have development approval for your home. Demolition usually takes only about a week and the builder will give you a couple of weeks to demolish while they order materials for the construction.

Construction

Construction starts. You will want to visit your home regularly as it is being built to see what is happening and note any possible problems. If you find any problems, talk to the builder about it immediately.

During construction, take as many photos as you can so that you can always locate pipes and wires later, although these will be shown on your plans if needed. Always ask questions to find out as much as possible about what will happen and when.

After handover, keep careful records of any problems for defect rectification. The builder will give you a maintenance form for a three-month maintenance review of your home. Note if doors are sticking or something needs to be tightened – all the little things that need to be adjusted. Of course, if you find something major then call the builder straight away, especially if its leaking taps or plumbing.

Useful contacts

National

Housing Industry Australia: <http://hia.com.au/>
Master Builders Australia: <www.masterbuilders.com.au>
To check if a builder or trade contractor is registered
 1300 360 320
 www.buildingcommission.com.au
Building Standards: Australian Building Codes Board
 1300 857 522
 www.abcb.gov.au
For general advice and assistance about building your home: Consumer Protection Advice Line
 1300 304 054
To check whether a trader is licensed
 Australian Securities and Investments Commission
 1300 300 630
 www.asic.gov.au
Taxation and Superannuation: Australian Taxation Office
 13 28 66
 www.ato.gov.au
 (If you build new residential premises for sale, you'll be liable for GST on the sale and entitled to claim GST credits for related purchases. If you renovate a property and sell it for a profit, there could be implications for income tax, capital gains tax and GST.)

ACT

Fair Trading: <www.ors.act.gov.au/FairTrading>

New South Wales

Building Advisory Service NSW: <www.buildingadvisory.com.au>

Northern Territory

Northern Territory Building Advisory Services Branch (BASB): <www.nt.gov.au/lans/building/index.shtml>

Northern Territory Consumer Affairs: <www.nt.gov.au>

Northern Territory Owner builder: <www.nt.gov.au/lands/building>

Queensland

Home Builders Advisory Service: <www.hbas.net.au>

Office of Fair Trading: <www.fairtrading.qld.gov.au>

Queensland Owner builder: <www.bsa.qld.gov.au>

South Australia

Building Advisory Services: <www.ocba.sa.gov.au>

Office of Consumer and Business Affairs: <www.ocba.sa.gov.au>

Owner builder: <www.ocba.sa.gov.au>

Tasmania

Building Advisory Services: <www.wst.tas.gov>

Consumer Affairs and Fair Trading: <www.consumer.tas.gov.au>

Owner builder: <www.wst.tas.gov.au/building>

For information about the Housing Indemnity Act 1992 contact the Office of Consumer Affairs and Fair Trading
1300 65 44 99
www.justice.tas.gov.au/newca/index.htm

Victoria

Building Advice and Conciliation Victoria
1300 557 559
www.buildingcommission.com.au/www/html/2540-complaints-disputes--appeals.asp

Consumer Affairs: <www.consumer.vic.gov.au>

Owner builder: <www.buildingcommission.com.au>

Western Australia

Western Australia owner builder: <www.builders.wa.gov.au>

Further information is available from the BDT publication: *Resolving Building Disputes – A Guide for Owners and Builders* available from <www.buildingdisputes.wa.gov.au>

To check whether a trade or business name is registered
Business Name Registration
1300 30 40 14
219 St Georges Terrace
Perth WA 6000
www.commerce.wa.gov.au

Advice on contract, legal obligations, technical queries and concerns
Housing Industry Association (HIA)
Consumer Line: 1902 973 555
22 Parkland Road
Herdsman Business Park
Osborne Park WA 6017
com.au/hia/region/WA.aspx

Advice on whether a builder is a member of the association and general advice to member's clients on building a new home
Master Builders' Association
9476 9800 or 1300 550 262 (country callers)
35 Havelock Street
West Perth WA 6005
mba@mbawa.com
www.mbawa.com

To check whether a builder is registered and information on how to lodge a complaint with the Building Disputes Tribunal
Builders' Registration Board
9476 1200
Suite 10
18 Harvest Terrace
Perth
www.builders.wa.gov.au

Building Disputes Tribunal Deals with building disputes
9476 1222
www.buildingdisputes.wa.gov.au

19

Project management

You have decided to owner-build rather than employ a contract builder, and by now you will have a pretty good idea of what is expected of you. There are lists as long as your arm of what you need to do to make a success of your project. Almost everything involved with your project hinges on your abilities.

We all start projects with good intentions, but a lot of would-be owner builders underestimate the amount of time required for their project. Also if you run into problems it can make you feel discouraged. The owner builder loan conditions state you must start the construction within 12 months of the loan and finish within two years. Although most owner builders think they will complete the construction within six to nine months, they are unlikely to succeed, as often not even project home builders with all their expertise can complete a home with this time frame.

Managing your development will not be an easy task. However, there are many resources at hand to aid you. The amount of money you save by owner-building will depend on how well you supervise the construction of your home, how much time and effort you devote, and how well you listen to good advice. Lost time, lost materials, stolen items, workers not showing up, the wrong workers showing up, late deliveries, wrong deliveries, not estimating the amount of materials properly and having to go out and replace lost materials are just a few of the problems that occur during the construction phase. If you rush through the planning stages, you will have trouble completing the project and will lose time and money in labour and materials during the construction. Preparation and planning will save you time, money and headaches.

To be successful in building your new home as an owner builder you will need to study every aspect of the construction process, and understand the timelines involved during the construction stages and how to manage each step accordingly. Even if things don't go exactly to plan – and there will undoubtedly be hiccups along the way– building your home will be one of the most satisfying and rewarding challenges you may ever undertake.

What management involves

The successful owner builder will be competent at planning. Most people make the mistake of believing that the most important part of project is the construction phase. However, the smart owner builder realises that it's the planning phase that is the maker or breaker.

For the owner builder, the construction process is, simply put, a very long project. You must have a wide range of organisational skills, because an owner builder is the general contractor for the construction. It will be your job to keep all of the subcontractors informed about the time line for the development. For instance, if your slab is taking longer than anticipated, you will need to let your framing team know the day before that they are not required. Otherwise they will show up for work and there won't be anything for them to do. You do not want to get them offside, as their time is money and often they are travelling some distance from home to work.

Since you will be hiring your own subcontractors and overseeing their work you must be able to successfully manage the hiring of multiple subcontractors as well as manage their labour during construction. It would be extremely valuable for you to learn what each of the trades do and study the different aspects of construction, as this will assist you in scheduling your subcontractors. As you will be allocating the tasks for the day you should be there in the morning to make sure they show up to work on time, stay on site to do a full day's work, and provide quality work on your home.

By the time your project is about to begin you should have decided on your suppliers, at least for all major items. As project manager you will be responsible for ordering, receiving, and inspecting deliveries to make sure the quantity and quality match your order, and for chasing refunds for any shortfalls or damage.

Ask your suppliers to advise you in advance of the date and time of any deliveries. Do not have items delivered before they can be installed as they could get damaged, stolen, or simply get in the way. Discuss your delivery arrangements with your trade contractors prior to engagement so that they know what to do if you are unable to be on site when a delivery is made. It is paramount you have the materials required at each stage of the development ready for the contractors to do their job. If the materials are not on hand for any reason, it is your task to reschedule contractors so you don't lose valuable time and money from lost productivity.

As each phase of construction is completed, you must make sure that you organise the inspections by the relevant building surveyors to the Building Code of Australia's requirements and to certify the workmanship and materials are done properly.

Engaging a project manager

What if there was a way of being an owner builder without having to do all the project management? Fortunately there are experienced professional project managers that you can engage.

Builders work on 20–30% margins, so if a home costs $300 000 to build the builder will make around $90 000. The project manager works on 5–7% of the cost for the construction, which is a much better proposal. Employing a project manager means you have someone there to do all the managing and leaves you to get on with the important stuff. You may only need help until lock-up stage; this is entirely up to you.

The following duty statement of a project manager demonstrates how their expertise will help you manage the construction.

- A project manager is responsible for looking after every detail of the project.
- The project manager has the responsibility of planning the project so that it is completed successfully, within the given deadline.
- The planning process involves forming a team of tradespeople required for effective completion of the project. The project manager will oversee the whole project and assign the trades to their specific roles. They will have excellent management skills to coordinate the services of everyone involved, and will also be able to work well with you.
- They should conduct regular meetings with tradespeople to keep up to date on the status of the project. It is the job of the project manager to supervise whether the tradespeople are working efficiently.
- They will make changes and improvements, if necessary, to achieve the desired results.
- The project manager will plan the resources needed for the project and chart a budget plan.
- The project manager will see to it that all the projects under their supervision go smoothly. They act as a link between all the parties and deal with problems that may arise within the project.

As the job of a project manager carries huge responsibilities, two most important things required of a project manager are planning and organisation skills. Along with these, the project manager will have a variety of other competencies:

- financial management skills
- good communication (verbal and written) and interpersonal skills
- excellent business management and developmental skills
- efficient team management skills
- the ability to resolve conflicts
- computer or technical knowledge.

Even project home builders employ a project manager to oversee their projects. Most project home builders are not master builders, and they hire a project manager to oversee the construction work and hire subcontractors.

When you engage a project manager, you will need to ascertain how many projects your project manager has and how much time they will be able to allocate for you. Some project managers for large builders will organise the tradespeople but will visit your site very seldom, and very briefly.

Computer assistance

If you decide to be your own project manager there are computer programs you can purchase online to help you manage your project better. These programs allow you to keep an electronic form like a database of all stages of the project. It's a smart and efficient use of your time and will save you money. The programs give you a step-by-step guide on planning and management that will give you an edge with controlling costs, time management and scheduling, and these all go together to provide an important management tool.

The information gathered for the program enables you to assess progress and performance of the project accurately against your budget expenditure. You can review and change the parameters in the event there are any unpredicted factors such as bad weather, late arrival on site of materials or trade contractors, and so on. It is a living document which will need regular revision.

If you do not wish to purchase a program you can set up a spreadsheet. It will take a bit of time and organisation on your part, but you can do it using the information in this book to guide you through the steps and processes that are needed.

20

Colours, concepts and common problems

To achieve the most satisfaction out of your new home or renovation you must take your likes and dislikes in character and appearance into account. Popular building styles include colonial, modern contemporary, charming country, and Tuscany. Many people want to include aspects of feng shui (a Chinese system for arranging our environment or surroundings to create harmony). Whatever your preferences, they demand considerable attention and planning.

If you find it difficult to choose a style or theme, explore home magazines, display homes, building showroom centres, furniture showrooms, kitchen and bathroom companies, wall and floor tile centres. Go to the places that will suit what you are trying to achieve; for example if you are looking for modern contemporary style or theme, you would find these styles only in a newly built and up-to-date display village.It is easier and cheaper to see a finished result than to speculate. If you still find it hard to make decisions on colours and concepts or fear making an expensive mistake, contact interior decorating consultants for assistance. People in this field can be found listed in your local advertising papers, Yellow Pages, and on the internet. Another source available is the Dulux Paint Colour Consultant, a company that can assist with interior and exterior colour selections and with colour effects such as dragging, sponging and rag rolling. A lot of country-style homes use these methods of painting on the wall linings to one metre in height from the floor, with a skirting as a border.

Tip: An excellent idea we got from a magazine is to have one row of floor tiles next to the skirting board and put carpet in the centre. It looks great.

Colour schemes

To help in getting started, colours that match well together are listed below. Use dark colours in moderation, as accents.

Colour box 1

- Navy blue, grey, peach, green, light green, dark purple, medium purple, lemon.

Colour box 1 is calming and mellow with the potential to make a room fresh again. This is one of the combinations to use in small areas and help soften the light. Good for rooms where people communicate, like dining rooms, family rooms and living rooms. Dark purple should be used only in small doses as it can be too heavy. Blue is peaceful. Green works well with natural colours and in baby rooms.

Colour box 2

- Black, dark grey, ivory, indigo, pale blue, white, light gold, light yellow, pale yellow.

Colour box 2 has a light and welcoming feel. It works well in kitchens and bathrooms with the introduction of brighter and bold cool colours. Black and white are opposites and balance each other. Light shades of gold, yellow and blue against black and white give feelings of calmness and distance.

Colour box 3

- Chocolate brown, medium rose pink, peach, medium grey, light grey, storm sky blue, black, off white, bright red.

Colour box 3 has warm and homely feelings from nature and earth. It works well as a base colour scheme and absorbs light and colour in moderation. It is ideal for accessories, throws, cushions, lounge suites and timber furniture. This colour combination is good for the study room and backdrops.

Colour box 4

- Peach, light peach, light yellow, matt yellow, medium tan, dark brown, orange, medium grey, light grey.

Colour box 4 has the potential to lift the spirit and gives warmth to a room. Use yellow in dark spaces and pale yellow in bedrooms. Pink is popular for female bedrooms. Use hot pink in kitchens and dining rooms as an accessory colour. Dusty pink suits country styles and themes. Orange is exciting, uplifting and gives energy for secondary colours.

Deciding on the paint colour

When deciding on the wall colours for our home, we used sample pots of paint instead of going ahead and painting the whole area and hoping for the best outcome. The light entering into a room can vary the colour tone on the walls, and a lot of light can make the colour look washed out. A dark room can make the colour appear darker.

We painted a piece of cardboard approximately 50 × 50 cm in size with two coats of paint from our sample pots (one colour per piece of cardboard) and when they were dry we placed them in the various rooms around the house at different times of day and at night, taking into consideration artificial (electric) light.

This enabled us to decide together if we liked the effects of each colour over a 24-hour cycle. Additionally, looking at the furniture and accessories in the home with the colour sample boards helped us with the final decision. We allowed ourselves over a week to settle on the range of colours to avoid putting ourselves under too much pressure.

Using these simple methods prevents unwelcome surprises and costly errors, and helps to identify a colour that looks good in every light.

Lessons from experience

Owner builders are generally creative people who have great energy and imagination. Therefore, when you have completed your home building, you will almost certainly start considering what you should and should not have done. You may come up with a better understanding of how you could have done something because of past experiences and because of wear and tear on the dwelling.

The following examples may help you avoid some pitfalls.

External and internal problem 1

When we were installing our hot water system, we found that the plumber had not tightened the plumbing underneath the kitchen sink properly. By the time we realised, 30 m^2 of the ground floor was flooded. Fortunately, we had timber skirtings and architraves and not MDF material, otherwise this job would have needed to be redone as the MDF would swell. The advice here is to check all plumbing connections before installing the hot water system.

External problem 2

The driveway we decided on was stamped dark grey concrete. We regret that we didn't choose medium grey pavers like our neighbour, as our driveway has faded while the neighbour's driveway still looks fresh and new.

External problem 3

The first house that we owner-built (and have since sold) has external bagging. The external bagging product we used has since been taken off the market because it had a short life span; on our old house it has discoloured and become patchy. The advice here is to purchase a good quality bagging or render that has been proven to have a long life span. The safest option is not to use new products in order to avoid unknown results.

External problem 4

We are very happy with the look and style of our Colorbond metal roof, but we hear the rain coming down on it at 3 a.m.. Additionally, there are cracking noises when the sun is rising and going down at the end of the day. The advice here is decide if you are sensitive to noise and can live with these extra sounds in your roof.

Internal problem 2

Down-lights in the roof cavity have been a problem ever since they were mixed with insulation. The transformers attached to the down-lights will overheat if they are covered by insulation, and the hot transformer can start a fire in the roof cavity. This happened to one of the display homes we worked at. The lights were meant to stay on overnight for better presentation of the home to passing traffic. Fortunately, the safety switch was triggered, turning off the transformer. This saved the house from burning down because the small fire in the roof went out.

Kitchen design problem

In the kitchen area we have black granite bench tops, and we have black granite vanity tops in the bathrooms. Even the slightest fingerprint is visible on the benches, and an ongoing need to wipe them off with a chamois when children and visitors lean on them. The advice here is to consider a bench top that is not high gloss or a dark colour.

Some design tips

Bathroom design

A clear pivot shower screen door has a tendency to drip water on the floor once opened after a shower, and to show soap stains. Stegbar Australia sells opaque frosted space-saver sliding shower screen doors that avoid these problems.

A sunny aspect for a bathroom is ideal for natural light in the wet areas, to dry condensation, and to make grooming in a mirror easier.

Design idea 1

One of the great ideas we picked out of a magazine and put into action was to lay one row of large floor tiles around the perimeter of the lounge and dining rooms, then lay carpet in the centre.

Design idea 2

Built-in 3D walls are a very popular and modern feature for a plasma home theatre set up, and are not very expensive.

Lounge rooms and home theatres should be placed away from sleeping areas for more peaceful surroundings. We placed insulated acoustic batts in the internal wall cavities around the bedrooms to provide more quiet.

Design idea 3

Outdoor living is a way to extend living areas and add more room for entertaining family and friends. Some single-storey houses are under 25 m^2 and feel so much bigger when an outdoor living area is adjacent to an indoor family room or rumpus. This is one option for those who are on a budget and limited in the size of home that they can build.

21

Gardens and landscaping

Back yard

Designing the back yard is just as important as designing the home. We are increasing looking to outdoor living and barbecue cooking, at the same time as back yards are getting smaller with reduced land size, larger houses and less private open space. Councils also develop control plans and guidelines that must be obeyed.

It is important to decide early what type of use the backyard will be put to. The best action to take would be to create a landscape plan and draw to scale. A professional landscape company could assist you with quotes and labour, but the more you learn to do yourself the cheaper it will be.

Before you commence the plan, it is essential to prepare for the future – electric power and lighting, barbecue gas points, speaker wiring, outdoor heating, external air conditioning unit and so on. Using rainwater from tanks for the laundry, toilet systems and gardens is highly recommended to save on water bills and environment resources. Start with designing the largest to smallest items in the plan. The following list may give some ideas.

Large items

- outbuildings – sheds, greenhouse, business office, granny flat, studio, aviary
- large play items – kids' swings, cubby house, trampoline
- lawn, secret garden, trees
- pergolas, decking

- side access for boats, caravans, cars
- retaining walls
- water tanks
- absorption beds in bushfire prone areas to help create a buffer zone.

Small items

- fruit and vegetable patches, garden beds
- clothesline
- ponds, water features
- dog kennels
- outdoor kitchen, barbecue, fire pit
- outdoor furniture, seating, hammocks
- storage sheds
- garden sculptures and decorative found objects such as milk cans
- steps
- pots, wine barrels
- privacy screens, wind breaks.

One very important component is the clothesline. It takes some thought to locate such a simple item, but it is crucial. If it is possible, a clothesline facing north would gain most of the sun during the day.

Front yard

Focus on creating the right front yard. There are so many opportunities and creative ideas available through media, display homes and magazines today that it would be easy to find something that suits your taste and desires.

A simple and elegant front garden does not dominate the façade of the home. Deciding on the main feature would be a start in planning the front yard. Is the attention to be focused on the home or something in the garden? Perhaps the front door is to be the main attraction, with a detailed concrete path to the street. An option may be a large circular garden in the middle with a beautiful flowering tree that holds its leaves for most of the year.

The possibilities for a garden are endless. You may instead want a bush garden to attract birds, or want a hedge to screen you from the street or protect against dust and noise, or you may have a passionate love of a particular species or type of plant.

Take a look around the neighbourhood and see what others have created for their gardens. What is the dominant theme in the streetscape? Maybe you want to blend in, or maybe you want to be totally different. Are there guidelines to follow for landscaping in new developed estates? Whatever your ideal, standing across

Stormwater

Stormwater drainage is an essential component of landscaping. It can affect not only your property, but the neighbours as well. When we were building the first home we had an unexpected flood in the back yard after a storm and heavy rain. The dwelling was up to roof stage, but still did not have the downpipes in place. Our neighbour's property to the right side of our dwelling was approximately 600 mm lower. We had plans to build a retaining wall once the house was completed, but with no downpipes and no retaining walls, this only made it easier for a flood to happen.

We discovered plastic temporary downpipes when building our second home, and invested in them as a precaution against flood. Companies that sell roofing products can assist with locating temporary downpipes.

the road opposite to the façade is a great way to help make decisions and allow your imagination.

Choosing plants

Choosing plants for the garden and climate can be a major issue. You need to analyse the time you have to spend in the garden watering and fertilising. Plants can be expensive, but many nurseries are able to help with selection and advice.The help of a professional horticulturalist could be worthwhile if you don't have the know-how with plants.

Take into consideration:

- the mature height and width of the plant
- maintenance
- pruning requirements.

The garden beds require drainage because plants and flowers can die if they are continually soaked in wet soil.

Garden materials that work well with plants are excellent themes and features. Some that come to mind as follows:

- coloured pebbles
- gravel
- stepping stones
- tiled walls
- timber for framing garden beds.

Using unusual items such as an aged fire extinguisher placed in a garden bed can create interest.

Appendix A: Resources

State authorities for consumers and traders

ACT: Office of Fair Trading: <www.ors.act.gov.au/FairTrading/index.html>
New South Wales: Department of Fair Trading: <www.fairtrading.nsw.gov.au>
Northern Territory: Consumer Affairs: <www.nt.gov.au>
Queensland: Office of Fair Trading: <www.fairtrading.qld.gov.au>
South Australia: Office of Consumer and Business Affairs: <www.ocba.sa.gov.au>
Tasmania: Office of Consumer Affairs and Fair Trading: <www.consumer.tas.gov.au>
Victoria: Consumer Affairs: <www.consumer.vic.gov.au>
Western Australia: Department of Commerce: <www.commerce.wa.gov.au>

Events and information support centres

Australian Handyman Magazine (available from Bunning's stores): <www.readersdigestadvertising.com.au>
Building Centre: www.buildingcentre.com
HIA Building & Renovating Expo Brisbane, Sydney, Melbourne and Adelaide:

- www.hiahomeshows.com.au/brisbane/index.htm
- www.sydneyhomeshow.com.au
- www.onlymelbourne.com.au
- www.hiahomeshows.com.au/adelaide/greensmart.htm

Home ideas: <www.homeideas.com.au>
'How to' channel (Foxtel) has broadcast an owner builder series, so watch for repeats or similar programs in future
Key 2 Owner Building: <www.key2ownerbuilding.com.au>
Practical Owner & Builder Magazine: <www.ownerbuild.com.au>

Renovate & Extend magazine: <www.isubscribe.com.au>
Build Your Own House: <www.byohouse.com.au>
The Owner Builder magazine: <www.theownerbuilder.com.au>

Owner Builder Magazine Article

Note: The following article was printed in *The Owner Builder* Magazine (June/July 2008).

We saved $80,000 by owner building

We are your average married couple, and have successfully owner built two project homes in the last six years. All you need is commonsense and a bit of patience.

My husband Matt and I had both worked in sales in the building industry, and didn't want to pay a large profit to a builder. We have learned all the tricks of the trade with big project builders. That's why we wanted to be in control of our own time and money; there is nothing worse when building a house than having someone else in control of YOUR time and money.

Keeping the mortgage down

If you have or can make the time, and apply the commonsense and patience required for owner building, you will save at least 20–30% on the builder profit. Not to mention the materials and labour – but this will now all be in your control. If you get an estimator to cost up your project and use it as a budget, then you're on the right track. It would be a bit risky going into a project without a realistic estimated budget.

We saved over $80,000 by owner building our second house, we know this because we had a builder quote on our design and project before we built it. It's crazy with the amount of money involved and the effect it has on your mortgage, especially with the way interest rates are going at the moment. I think owner building speaks for itself: save where you can, your future depends on what you spend today.

Paperwork and council approval

When we purchased our land there was a height restriction covenant on it. That meant we could only build a single storey house with a maximum of 5.6 metres to the highest point of the roof. We had planned to rebuild a replica of our first house again, which was a single storey with two lofts in the roof and plenty of storage space. Therefore, we had to redesign the plans to suit. Once finalised, these were taken to a building designer to create the working drawings from our original concept.

To gain a bit more height space we excavated further into the ground and therefore had to pay more to remove the excess spoil. It was worth it because we like to have high roof pitches on our homes, it just adds more character.

We were lucky enough to have a building consultancy service to organise all our paperwork and approvals with the council. They do exist and they are a big help. We paid $350 for the service and it was worth every cent. The company would call us when the next step was ready to start and tell us what was required. To find such a service for your project I would suggest you search for building consultants on the Internet, the trade section in local newspaper, and the Yellow Pages.

As we all know, council inspections are necessary. When we received our council approved plans, the attached paperwork stated the required inspection stages during construction. We used an accredited private certifier in lieu of a council inspector, which is usually a faster service then using council. It also means that you can use someone who has specific knowledge of the type of construction you are undertaking, where this may be an issue (such as earth or straw bale building). Fortunately all our inspections passed with no problems.

Building materials

We built using a raft concrete slab, timber frame, fired clay brick veneer and *Colorbond* roof. The house, verandahs and garage cover 350m^2.

We chose the raft slab as it is cost effective and flexible, allowing for ground movement that helps prevent multiple cracking in brickwork and cornices.

When our slab edge boards were in situ and the plumber was preparing the internal drainage, there was a near disaster. This is where the commonsense part I mentioned comes in. Our plans had included a one metre wide verandah on the back of the house. The concreter had correctly placed the edge boards for the perimeter of the slab, at the edge of the verandah area. However, the plumber judged his measurements for the plumbing from the perimeter of the verandah and not the perimeter of the house. Luckily Matt noticed the error before the slab was due to be poured the next day. I had felt something was wrong when viewing the size of the family room, adjacent to the rear verandah; it felt too small, seeing as the kitchen plumbing was next to the family room. If this error hadn't been picked up, our main bedroom and family room would have been short by a metre each.

Timber frames are easier to plane and adjust for plasterboard internal linings. When our frame was being erected we were very excited – to see each room and walk in and out of them is so rewarding. Our plans were coming to life after all the preparation and waiting. There were no dramas with the frame or framer and it was one of the best moments.

Fired clay brick has that solid feeling about it and looks attractive as well. We purchased 'commons' from the brick supplier, due to the fact we were planning to bag the brick. Matt then bagged the whole perimeter of the house himself, and he must have lost some kilos doing it. He did have a lot of spare time on the weekends to do it, while I worked selling houses off the plan for a builder. He would joke with me in the morning and say, 'Don't come home unless you sell one;' I would reply, 'Just worry about building our house and finishing it.'

A *Colorbond* roof has a certain style about it. Have you ever heard someone say they love the sound of rain on a tin roof? Well I am one of them. You will need a roofer to install your tin roof, preferably one that has a good reputation and workmanship. On our first house we had some problems with the roof. Everything was going fine until the company wrote to us saying they would not do any further work until we paid our account. This was confusing because we were up to date, and proved it with our cheque book. It transpired that the salesman we dealt with had a gambling problem, and had cashed in the cheques we gave him for payment on the roof and labour. Luckily it was sorted out and we were not affected by the incident financially.

The approximate cost breakdown to this stage was:

Slab	$27,000 (includes piers and labour)
Frame	$16,000
Brick supply	$6,500
Roof	$20,000 (includes labour)

Heating, cooling and insulation

In both of our homes we installed ducted airconditioning and insulation. The insulation was installed in the ceiling, some internal walls (bedrooms), and between the frame and brickwork of all external walls. There is also extra heavy duty *Gladiator* wall wrap insulation around the external frame. This all helps with noise, spiders, and heating and cooling the dwelling. Our neighbours say they never hear us, even though we have two young babies – we think it might have something to do with the insulation. Discuss your requirements with an insulation supplier.

Finishing

Builders usually have a standard inclusion list for their homes. If you want the good gear, then you have to pay more. When you owner build, you are able to shop around at different suppliers to obtain the best deal for each inclusion.

For example, some of the 'upgrades' we made were: wide skirtings and architraves; decorative cornices; granite benchtops in kitchen and bathroom, with a glass splashback on the kitchen wall; European kitchen appliances and quality kitchen cabinets; quality tapware, toilets, basins and bath.

This is also an area that can make or break the budget, so don't discount the potential for savings here.

Managing the project

While we used tradespeople for a lot of the more specialist tasks, we also got in there and got our hands dirty! After all, isn't that all part of being an owner builder? We did the site clean up, painting, bagging, installation of airconditioning ducts and insulation, as well as acting as general labourers.

Dealing with contractors was generally fairly easy, as long as we were clear right from the start about the outcomes of each stage. Any ambiguity would escalate into a problem unless well managed.

Other major items included: airconditioning $7000, insulation $1500, windows $6000, doors $3500.

How much can you save?

How long is a piece of string? To start with, you can at least save the 20–30% profit that would be spent if hiring a builder. Even if you hire a project manager, you still will save thousands of dollars. How much more you save is very dependent on how much you do yourself, taking into account lost earnings should you decide to give up a paid job to complete the project. Although you may miss out on some of the bulk savings made by large builders, you will gain on the reduction of waste that is inevitable on large projects. On top of this, you will also get a house that exactly meets your needs, rather than having to compromise because the option you wanted was not available.

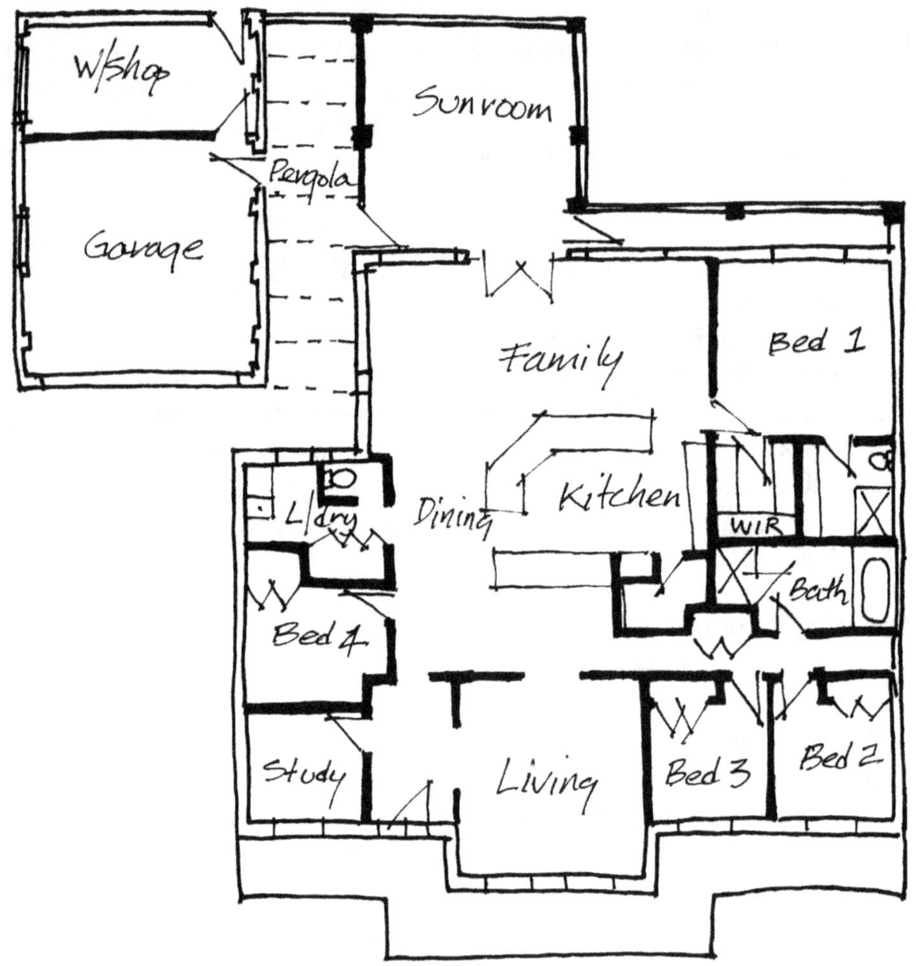

Gladiator

Extra heavy duty wall wrap insulation manufactured from aluminium foil, laminated to a high strength reinforcing polymer weave. *1300 300 249, www.insulationsolutions.com.au*

Boral Bricks

A wide range of bricks, colours, shapes and textures.
13 30 35, www.boral.com.au

Stegbar

Highly crafted windows, doors, showerscreens, splashbacks and wardrobes.
1800 681 168, www.stegbar.com.au

Hume Doors & Timber

Extensive range of both internal and external timber doors, door skins and timber mouldings.
02 9794 1111, www.humedoors.com.au

Appendix B: Trades, materials and services directory

Hiring of tradespeople and businesses and purchasing materials are the sole responsibility of the customer. As the decisions, actions and outcomes will be at the customer's own risk, they should use the state authority services available.

We encourage checking all licensing, registration and businesses before committing to a business, as discussed in Chapters 2 and 8.

There are many possible suppliers for some items so this list cannot be exhaustive.

Directory in 68 steps

This directory corresponds to steps 1–68 in Chapter 4, 'The Building Process in 68 Steps'.

1. Contour levels and geotechnical engineer

www.startlocal.com.au
www.homeimprovementpages.com.au
www.besttradebid.com.au
www.tradeconnect.com.au
www.ghd.com.au
www.outdoordesign.com.au

2. Design and cost your house

Design

www.homebuildersaustralia.com.au
www.tradeconnect.com.au
www.homeimprovementpages.com.au
www.besttradebid.com.au
www.etradesman.com.au

Estimator

www.tradeconnect.com.au
www.pdhestimating.com.au

3. Owner builder course

Check with your state authority for new accredited owner builder course providers.

4. Building permit

ACT: Office of Fair Trading <www.accc.gov.au>
New South Wales: Department of Fair Trading <www.fairtrading.nsw.gov.au>
Northern Territory: Consumer Affairs <www.consumeraffairs.nt.gov.au>
Queensland: Office of Fair Trade <www.fairtrading.qld.gov.au>
South Australia: Office of Consumer & Business Affairs <www.sa.gov.au/tags/building%20permits>
Tasmania: Office of Consumer Affairs & Fair Trading <www.consumer.tas.gov.au>
Victoria: Consumer Affairs Victoria <www.consumer.vic.gov.au>
Western Australia: Department of Commerce <www.docep.wa.gov.au>

5. White card course

www.ownerbuilding.com.au

6. Finance

www.homebuildersaustralia.com.au
www.besttradebid.com.au
www.startlocal.com.au
www.mortgagechoice.com.au

7. Engineer for slab construction

www.startlocal.com.au
www.homeimprovementpages.com.au
www.besttradebid.com.au
www.tradeconnect.com.au

8. Sustainability standards

www.basixconsulting.com.au (New South Wales)
www.ecoengineering.com.au
www.ecocertificates.com.au

9. Developer application

The company you purchased the land from.

10. Council approval and colour selections

The council your land is zoned under.

11. Housing Industry Association

www.hia.com.au

12. Water authority approval

The last invoice bill for water service or council will tell you.

13. Owner builder insurance broker

www.buildersbroker.com.au
www.buildsafe.rtrk.com.au

14. Certificate to commence construction

Your local council.

Private certifier or building consultant

www.startlocal.com.au
www.homeimprovementpages.com.au
www.besttradebid.com.au
www.tradeconnect.com.au

15. Retaining walls

www.tradeiefinder.com
www.localtradesmendirectory.com
www.homeimprovementpages.com.au
www.besttradebid.com.au

Trench for retaining wall

www.homeimprovementpages.com.au
www.homebuildersaustralia.com.au

To purchase footings for the retaining wall; concrete

www.besttradebid.com.au
www.buybuildingsupplies.com.au
www.buildingtradesupplies.com.au
www.homeimprovementpages.com.au
www.homebuildersaustralia.com.au
www.ibs.com.au
www.bunnings.com.au

To pump concrete into footing of retaining wall

www.besttradebid.com.au
www.homebuildersaustralia.com.au

16. Site toilet

www.outdoordesign.com.au
www.hotfrog.com.au

17. Peg-out or survey

www.homeimprovementpages.com.au
www.startlocal.com.au
www.tradeconnect.com.au

18. Excavator

www.truelocal.com.au
www.homebuildersaustralia.com.au
www.homeimprovementpages.com.au
www.etradesman.com.au
www.tradeconnect.com.au

19. Site signage

www.tradeconnect.com.au
www.truelocal.com.au

20. Site access

www.hotfrog.com.au
www.truelocal.com.au
www.besttradebid.com.au

21. Organising materials and contractors

Frame and trusses

www.besttradebid.com.au
www.hotfrog.com.au
www.yellowbook.com.au
www.homeimprovementpages.com.au

Framer

www.tradeconnect.com.au
www.yellowbook.com.au

Windows

www.buybuildingsupplies.com.au
www.stegbar.com.au

Glass bricks

www.tradeconnect.com.au
www.buybuildingsupplies.com.au

Brick supply and steel

www.ibs.com.au
www.buybuildingsupplies.com.au
www.buildingtradesupplies.com.au
www.besttradebid.com.au

Bricklayer

www.localtradesmendirectory.com
www.homebuildersaustralia.com.au
www.etradesman.com.au
www.homeimprovementpages.com.au
www.tradeconnect.com.au
www.tradiefinder.com
www.besttradebid.com.au

Roof supply

www.buybuildingsupplies.com.au
www.tradeconnect.com.au
www.tradiefinder.com
www.metroll.com.au
www.buildingtradesupplies.com.au
www.besttradebid.com.au
www.ibs.com.au

Scaffolding

www.homeimprovementpages.com.au
www.startlocal.com.au
www.tradeconnect.com.au

Metal fascia and gutter

www.metroll.com.au
www.getprice.com.au
www.stratco.com.au

www.buybuildingsupplies.com.au
www.buildingtradesupplies.com.au

Bull nose roofing

www.buybuildingsupplies.com.au
www.stratco.com.au

Roofer

www.serviceseeking.com.au
www.tradeconnect.com.au
www.etrademan.com.au
www.besttradebid.com.au
www.etradesman.com.au

22. Piers

www.besttradebid.com.au
www.servicecentral.com.au
www.homeimprovementpages.com.au
www.etradesman.com.au
www.homebuildersaustralia.com.au
www.tradiefinder.com
www.tradeconnect.com.au

23. Forming the slab with concreter and plumber

www.homeimprovementpages.com.au
www.servicecentral.com.au
www.besttradebid.com.au
www.homebuildersaustralia.com.au
www.tradiefinder.com
www.tradeconnect.com.au
www.etradesman.com.au

24. Pest control

www.homeimprovementpages.com.au
www.servicecentral.com.au
www.besttradebid.com.au
www.etradesman.com.au
www.tradeconnect.com.au

25. Pour the slab

www.servicecentral.com.au

www.homebuildersaustralia.com.au
www.besttradebid.com.au
www.homeimprovementpages.com.au
www.etradesman.com.au
www.tradiefinder.com
www.tradeconnect.com.au

26. Excavation and connection of services

www.tradeconnect.com.au
www.etradesman.com.au
www.homebuildersaustralia.com.au
www.homeimprovementpages.com.au
www.servicecentral.com.au
www.besttradebid.com.au

27. Temporary fencing

Welded mesh, star pickets and labour

www.homeimprovementpages.com.au
www.buildingtradesupplies.com.au
www.besttradebid.com.au
www.bunnings.com.au
www.getprice.com.au

28. Erect the frame

www.yellowbook.com.au
www.tradeconnect.com.au

29. Electricity

www.tradeconnect.com.au
www.etradesman.com.au
www.homebuildersaustralia.com.au
www.tradiefinder.com
www.homeimprovementpages.com.au
www.servicecentral.com.au
www.besttradebid.com.au
www.localtradesmendirectory.com

30. Plumber and gas rough-in

www.tradeconnect.com.au
www.etradesman.com.au
www.homebuildersaustralia.com.au

www.tradiefinder.com
www.homeimprovementpages.com.au
www.servicecentral.com.au
www.besttradebid.com.au
www.localtradesmendirectory.com
www.servicecentral.com.au

31. Water tank

www.stratco.com.au
www.getprice.com.au

32. Bricklayer

www.tradeconnect.com.au
www.homebuildersaustralia.com.au
www.tradiefinder.com
www.homeimprovementpages.com.au
www.besttradebid.com.au
www.etradesman.com.au
www.localtradesmendirectory.com

33. Metal fascia and gutter and roof installation

www.buildingtradesupplies.com.au
www.ibs.com.au

34. Planing the frame and nail-off brick ties

www.yellowbook.com.au
www.tradeconnect.com.au

35. Council inspection

www.startlocal.com.au
www.homeimprovementpages.com.au
www.besttradebid.com.au
www.tradiefinder.com
www.etradesman.com.au

36. Clean site

www.etradesman.com.au
www.besttradebid.com.au
www.homeimprovementpages.com.au

37. Eave carpenter

www.homebuildersaustralia.com.au

www.tradiefinder.com
www.homeimprovementpages.com.au
www.localtradesmendirectory.com

38. Electrical rough-in

www.localtradesmendirectory.com
www.homeimprovementpages.com.au
www.tradeconnect.com.au
www.etradesman.com.au
www.homebuildersaustralia.com.au
www.tradiefinder.com
www.servicecentral.com.au
www.besttradebid.com.au

Products (retail)

www.aussieweb.com.au
www.besttradebid.com.au
www.localtradesmendirectory.com
www.tradeconnect.com.au

Supplies (wholesale: may require a trade account)

www.electricalwholesales.com.au
www.besttradebid.com.au
www.localtradesmendirectory.com

39. Brick cleaner

www.homeimprovementpages.com.au
www.tradeconnect.com.au

40. Wall insulation

www.localtradesmendirectory.com
www.tradeconnect.com.au
www.etradesman.com.au
www.besttradebid.com.au
www.ibs.com.au
www.buildingtradesupplies.com.au

41. Plasterboard internal linings

www.localtradesmendirectory.com
www.etradesman.com.au
www.besttradebid.com.au

www.ibs.com.au
www.buildingtradesupplies.com
www.tradiefinder.com
www.homebuildersaustralia.com.au
www.homeimprovementpages.com.au

42. Metal downpipes

www.metroll.com.au
www.getprice.com.au
www.stratco.com.au
www.buybuildingsupplies.com.au
www.buildingtradesupplies.com.au

43. Staircase and garage door

www.homeimprovementpages.com.au
www.truelocal.com.au

Garage doors

www.bnd.com.au
www.besttradebid.com.au

44. Install hobs for shower and bath

www.localtradesmendirectory.com
www.etradesman.com.au
www.besttradebid.com.au
www.tradiefinder.com
www.homebuildersaustralia.com.au
www.servicecentral.com.au
www.tradeconnect.com.au

45. Waterproof flashing in wet areas

www.homeimprovementpages.com.au
www.servicecentral.com.au
www.besttradebid.com.au

46. Council inspection

www.startlocal.com.au
www.homeimprovementpages.com.au
www.besttradebid.com.au
www.tradiefinder.com
www.etradesman.com.au

47. Final fix-out – carpenter

www.tradeconnect.com.au
www.homeimprovementpages.com.au
www.servicecentral.com.au
www.homebuildersaustralia.com.au
www.localtradesmendirectory.com

Materials

www.tradeconnent.com.au
www.buybuildingsupplies.com.au
www.humedoors.com.au
www.stegbar.com.au
www.getprice.com.au
www.hotfrog.com.au
www.etradesman.com.au

48. Install bath

Material

www.besttradebid.com.au
http://harveynorman.findnearest.com.au

Trades

www.localtradesmendirectory.com
www.etradesman.com.au
www.besttradebid.com.au
www.tradiefinder.com
www.homebuildersaustralia.com.au
www.servicecentral.com.au
www.tradeconnect.com.au

49. Tile wet areas

www.localtradesmendirectory.com
www.etradesman.com.au
www.besttradebid.com.au
www.tradiefinder.com
www.homebuildersaustralia.com.au
www.servicecentral.com.au
www.tradeconnect.com.au

Supply

www.atozpages.com.au

www.yellowpages.com.au

50. Install toilets and shower screens

Supply

http://harveynorman.findnearest.com.au
www.besttradebid.com.au

Plumbing trades

www.tradeconnect.com.au
www.etradesman.com.au
www.homebuildersaustralia.com.au
www.homeimprovementpages.com.au
www.servicecentral.com.au
www.besttradebid.com.au

51. Install kitchen, vanities and sinks

Kitchen carpenter

www.tradeconnect.com.au
www.etradesman.com.au
www.servicecentral.com.au
www.besttradebid.com.au
www.startlocal.com.au
www.harveynormanrenovations.com.au
www.localtradesmendirectory.com
www.tradiefinder.com

Granite

www.yellowpages.com.au
www.besttradebid.com.au

Vanity bowls supply

http://harveynorman.findnearest.com.au
www.tradelink.com.au

52. Plumber finish-off

www.tradeconnect.com.au
www.etradesman.com.au
www.homebuildersaustralia.com.au
www.homeimprovementpages.com.au
www.servicecentral.com.au
www.besttradebid.com.au

Supply tap, toilet, sinks

www.tradelink.com.au
http://harveynorman.findnearest.com.au
www.bunnings.com.au

53. Painting

www.etradesman.com.au
www.homeimprovementpages.com.au
www.servicecentral.com.au
www.besttradebid.com.au
www.localtradesmendirectory.com

54. Electrical fix-out

www.tradeconnect.com.au
www.etradesman.com.au
www.homebuildersaustralia.com.au
www.tradiefinder.com
www.homeimprovementpages.com.au
www.servicecentral.com.au
www.besttradebid.com.au
www.localtradesmendirectory.com

55. Floor – sand slab

www.localtradesmendirectory.com
www.besttradebid.com.au
www.servicecentral.com.au
www.tradeconnect.com.au

56. Clean site

www.tradeconnect.com.au
www.etradesman.com.au
www.homeimprovementpages.com.au
www.besttradebid.com.au

57. Floor coverings

Floor tiles

www.atozpages.com.au
www.besttradebid.com.au

Carpets

www.carpetone.com.au
www.hotfrog.com.au
www.au.ask.com
www.besttradebid.com.au

Timber floors

www.getprice.com.au
www.tradeconnect.com.au

Tiling and waterproofing

www.localtradesmendirectory.com
www.etradesman.com.au
www.besttradebid.com.au
www.tradiefinder.com
www.homebuildersaustralia.com.au
www.servicecentral.com.au
www.tradeconnect.com.au

58. Final pest control

www.servicecentral.com.au
www.tradeconnect.com.au
www.homeimprovementpages.com.au
www.besttradebid.com.au
www.etradesman.com.au

59. Driveway and paths

www.besttradebid.com.au
www.servicecentral.com.au
www.homeimprovementpages.com.au
www.etradesman.com.au
www.homebuildersaustralia.com.au
www.tradiefinder.com
www.tradeconnect.com.au

60. Roof insulation

www.etradesman.com.au
www.buildingtradesupplies.com.au
www.servicecentral.com.au

61. Shower screens and mirrors

www.stegbar.com.au
http://harveynorman.findnearest.com.au

62. Final council inspection

www.startlocal.com.au
www.homeimprovementpages.com.au
www.besttradebid.com.au

63. Professional clean of house

We did our own house cleaning.
www.etradesman.com.au
www.homeimprovementpages.com.au

64. Kitchen appliances and hot water system

www.appliancesonline.com.au
www.graysonline.com
http://harveynorman.findnearest.com.au
www.tradeconnect.com.au
www.getprice.com.au
www.startlocal.com.au
www.rinnai.com.au

Plumbing

www.tradeconnect.com.au
www.etradesman.com.au
www.homebuildersaustralia.com.au
www.homeimprovementpages.com.au
www.servicecentral.com.au
www.besttradebid.com.au

65. Blinds and light fittings

www.blinds-australia.com.au
www.blindswarehouse.com.au
www.halfpriceblinds.com.au
www.tradeconnect.com.au

66. Fencing and retaining walls

www.besttradebid.com.au
www.truelocal.com.au

www.etradesman.com.au
www.tradeconnect.com.au

67. External accessories

www.bunnings.com.au

68. Laying turf

www.sydneylawnandturf.com.au
www.palmetto.com.au
www.sirwalter.com.au
www.besttradebid.com.au
www.etradesman.com.au
www.tradeconnect.com.au

Appendix C: Budget planning worksheet

Item	Quote 1	Quote 2	Quote 3	Labour	Budget provisional allowance	Price paid
Finances						
Pre-approval of loan						
Deposit required						
Stamp duty on land						
Stamp duty on mortgage						
Mortgage insurance						
Loan application						
Solicitor costs						
Total:						
Preparation of site costs						
Building permit						
White card course						
Demolition cost						
Demolition council fee						
Draftsman or architect						
Services: power, water, sewer and stormwater						
Contour levels						
Geotechnical engineer						
Council inspector or private certifier						
Sustainability certificates						
Clearing of site						
Double handling of materials						
Insurances with owner builder broker						

Item	Quote 1	Quote 2	Quote 3	Labour	Budget provisional allowance	Price paid
Owner builder course (if required)						
Developer application						
Surveyor						
Water board						
Hydraulics engineer (if required)						
Council additional DA approved essentials						
Total:						
Excavation						
Cut and fill excavation				$		
Site toilet						
Signage						
Retaining walls						
Peg-out survey						
Site access (blue metal)						
Sediment control barrier						
Excavation for drainage						
Removal of additional soil						
Total:						
Slab or foundation				$		
Concrete materials						
Additional piering						
Pest control						
Plumber (drainage)						
Temporary fencing						
Additional concrete items: paths, steps and landings						
Driveway, crossover of nature strip						
Total:						
Frame and trusses						
Frame, trusses, battens materials						
Hockey sticks for verandah and timber poles (if required)						
Frame carpenter				$		

Item	Quote 1	Quote 2	Quote 3	Labour	Budget provisional allowance	Price paid
Frame hardware						
Eave material						
Windows and glass sliding doors						
Glass bricks (if required)						
Plumber and gas rough-in						
Gas company fee						
Electricity box						
Total:						
Bricklayer				$		
Brick supply						
Brick hardware: lintels, brick ties, bycol and oxide						
Structural steel						
Sand and cement						
Flashings and damp proof course						
Vents						
External brick coatings (bagging)						
Brick scaffolding (2-storey)						
Brick cleaner (if required)						
Total:						
Roof				$		
Metal fascia and gutter						
Roof material						
Verandah material						
Safety rail for roof						
Eave carpenter				$		
Electrical rough-in						
Cleaning of site and rubbish removal						
Pergola material						
Downpipes						
Total:						
Fix-out						

Item	Quote 1	Quote 2	Quote 3	Labour	Budget provisional allowance	Price paid
Council or private certifier Inspection						
Internal wall and cornices						
Wall insulation						
Ceiling linings						
Carpenter fix-out				$		
Materials for fix-out						
Doors and jambs						
Architraves and skirtings						
Robes, pantry, linen cupboards materials						
Door handles and knobs						
Air conditioning or Heating						
Garage door and automatic opener						
Staircase, balustrade and handrails						
Steps						
Total:						
Practical completion						
Waterproof flashing in wet areas						
Council or private certifier inspection				$		
Install toilets and mirrors				$		
Bathroom vanities, laundry cupboards						
Kitchen cupboards						
Kitchen sink, benches and splash back or tiles						
Wall and floor tiling				$		
Painting				$		
Plumbing materials and finish off						
Electrical finish off				$		
Install bath and hob, laundry tub, basins and shower screens						

Item	Quote 1	Quote 2	Quote 3	Labour	Budget provisional allowance	Price paid
Floor – sand slab 9if concrete)						
Floor coverings						
Blinds or curtains						
Toilet roll holders, soap dishes, towel rails and fixtures						
Tapware						
Roof installation						
Cleaning of site and rubbish removal						
Kitchen appliances: rangehood, fridge, dishwasher,oven and cooktop						
Exhaust fans and heat lamps						
Security alarm						
TV points and gas points						
Fly screens						
Total:						
Other costs						
Fencing						
Hot water system						
Water tank						
Turf						
Ducted vacuum system						
Removalist						
New furniture						
Letterbox and house number						
Clothesline						
Transfer of services: phone wiring, electricity, gas etc.						
Intercom system						
Gates						
TV antennae and connection						
Garbage disposal bins from council						

Item	Quote 1	Quote 2	Quote 3	Labour	Budget provisional allowance	Price paid
Door bell						
Security doors						
External taps						
Total:						

Appendix D: Endnotes

Chapter 1

1 www.nt.gov.au
2 www.ownerbuild.com.au
3 www.nt.gov.au
4 www.wst.tas.gov.au
5 www.circularhead.tas.gov.au

Chapter 5

1 www.howtobuildahouse.com.au
2 www.howtobuildahouse.com.au
3 Based on the Preface in AS 2870-2011. Reprinted with permission of SAI Global Ltd. Copies of the Standard can be purchased online at www.saiglobal.com.
4 Based on the Preface in AS 2870-2011. Reprinted with permission of SAI Global Ltd. Copies of the Standard can be purchased online at www.saiglobal.com.
5 toolboxes.flexiblelearning.net.au
6 www.renovationrobot.com.au
7 toolboxes.flexiblelearning.net.au
8 www.renovationrobot.com.au

Chapter 6

1 www.yourhome.gov.au
2 www.yourhome.gov.au
3 www.yourhome.gov.au

4 www.yourhome.gov.au
5 www.climatechange.gov.au
6 www.yourhome.gov.au
7 www.yourhome.gov.au
8 www.yourhome.gov.au
9 www.yourhome.gov.au
10 www.yourhome.gov.au
11 www.yourhome.gov.au
12 www.yourhome.gov.au
13 www.yourhome.gov.au
14 www.yourhome.gov.au
15 www.yourhome.gov.au
16 www.yourhome.gov.au
17 www.yourhome.gov.au
18 www.yourhome.gov.au

Chapter 7

1 www.builders-directory.com.au/understanding-house-structure.shtml
2 www.builders-directory.com.au/understanding-house-structure.shtml
3 www.builders-directory.com.au/understanding-house-structure.shtml
4 www.builders-directory.com.au/understanding-house-structure.shtml
5 www.builders-directory.com.au/understanding-house-structure.shtml
6 www.builders-directory.com.au/understanding-house-structure.shtml
7 www.builders-directory.com.au/understanding-house-structure.shtml
8 www.choiceconcreting.com.au
9 www.yourhome.gov.au
10 www.yourhome.gov.au
11 www.homeimprovementpages.com.au
12 www.homeimprovementpages.com.au
13 www.yourhome.gov.au
14 www.yourhome.gov.au
15 www.yourhome.gov.au
16 www.yourhome.gov.au
17 www.yourhome.gov.au
18 www.yourhome.gov.au
19 www.tastimber.tas.gov.au
20 www.cemintel.com.au
21 www.aip.gov.au
22 www.thinkbrick.com.au
23 www.thinkbrick.com.au

24 www.hebelaustralia.com.au
25 www.yourhome.gov.au

Chapter 11

1 www.abcb.gov.au
2 www.coag.gov.au
3 www.nathers.gov.au
4 www.nathers.gov.au
5 www.nathers.gov.au
6 www.hearne.com.au/products/accurate
7 www.solarlogic.com.au/bers-pro/details
8 www.nabers.com.au
9 www.basix.nsw.gov.au
10 www.sustainability.vic.gov
11 www.nathers.gov.au
12 www.abcb.gov.au
13 www.dlp.qld.gov.au
14 www.abcb.gov.au
15 www.absa.net.au
16 www.nt.gov.au
17 www.nt.gov.au

Chapter 12

1 www.rfs.nsw.gov.au
2 Reproduced with permission from The Rural Fire Service of NSW; published by The Rural Fire Service NSW Government Australia, 22/09/2010.
3 Reproduced with permission from The Rural Fire Service of NSW; published by The Rural Fire Service NSW Government Australia, 22/09/2010.
4 Reproduced with permission from The Rural Fire Service of NSW; published by The Rural Fire Service NSW Government Australia, 22/09/2010.
5 Reproduced with permission from The Rural Fire Service of NSW; published by The Rural Fire Service NSW Government Australia, 22/09/2010.

Chapter 14

1 www.treasury.gov.au
2 Reproduced with permission by NSW Fair Trading, 2010.
3 Reproduced with permission by NSW Fair Trading, 2010.
4 Reproduced with permission by NSW Fair Trading, 2010.

5 Reproduced with permission by NSW Fair Trading, 2010.
6 Reproduced with permission by NSW Fair Trading, 2010.
7 Reproduced with permission by NSW Fair Trading, 2010.

Chapter 15

1 www.planning.nsw.gov.au
2 www.planning.nsw.gov.au
3 www.planning.nsw.gov.au
4 www.planning.nsw.gov.au
5 Reproduced with permission from the NSW Government Department of Planning.
6 Reproduced with permission from the NSW Government Department of Planning.
7 Reproduced with permission from the NSW Government Department of Planning.
8 Reproduced with permission from the NSW Government Department of Planning.

Chapter 17

1 www.planning.nsw.gov.au
2 www.aibs.com.au
3 www.aibs.com.au

Index

www.ingramcontent.com/pod-product-compliance
Lightning Source LLC
LaVergne TN
LVHW061221100826
845148LV00004B/817

* 9 7 8 0 6 4 3 1 0 0 4 2 8 *